David A. Dwyer

Amos and the
Davidic Empire

Amos and the Davidic Empire

A SOCIO-HISTORICAL APPROACH

Max E. Polley

New York   Oxford
OXFORD UNIVERSITY PRESS
1989

Oxford University Press

Oxford   New York   Toronto
Delhi   Bombay   Calcutta   Madras   Karachi
Petaling Jaya   Singapore   Hong Kong   Tokyo
Nairobi   Dar es Salaam   Cape Town
Melbourne   Auckland

and associated companies in
Berlin   Ibadan

Copyright © 1989 by Max E. Polley

Published by Oxford University Press, Inc.,
200 Madison Avenue, New York, New York 10016

Oxford is a registered trademark of Oxford University Press

Library of Congress Cataloging-in-Publication Data
Polley, Max E.
Amos and the Davidic Empire : a socio-historical approach /
Max E. Polley.
p.   cm.
Bibliography: p.
Includes index.
ISBN 0–19–505478–4
1. Bible. O.T. Amos—Criticism, interpretation, etc.
2. Religion and politics—Biblical teaching.
3. Palestine—Religion.   I. Title.
BS1585.2.P65 1989
224'.8067—dc19   88–17484 CIP

2 4 6 8 9 7 5 3 1

Printed in the United States of America
on acid-free paper

For
Jackie, Vance, and Lynn

Preface

IN THE FALL of 1974, the late Dr. G. Henton Davies, Principal Emeritus of Regent's Park College, Oxford University, lectured in my Hebrew Prophets course at Davidson College. He chose Amos as his subject, focusing student attention on one verse, namely, Amos 1:2:

> From Zion the Lord roars,
> from Jerusalem he utters his voice.
> (Davies' translation)

Suppose, he argued, contrary to most scholarly opinion, that verse is an authentic Amos oracle.

I found his position intriguing. Thus began my study of the Book of Amos. My research has led me to a handful of scholars who share this approach to Amos, and they have all been duly acknowledged. But my deepest gratitude goes to Dr. Davies, friend and scholar, who encouraged and guided me along the way. My only regret is that he did not live to see this book in print.

This study is written for college and seminary students, ministers, and interested lay persons. Although it is not written for the specialists in Amos studies, I believe its distinctive approach will be of interest to students of Amos at all levels.

The book has grown out of teaching Amos to undergraduate students. I have organized it the way I teach, introducing information where it will best be understood. For example, Chapters 2 and 3 present essential background material for interpreting Amos within his socio-historical setting. In Chapter 3, I discuss the administration of the state and the nature of state religion under David and Solomon in preparation for understanding Amos' attack on state religion in northern Israel. An analysis of Amos' oracles does not begin until Chapter 4. Although this delays the development of the main thesis, I frequently relate the background material to Amos. The alternative is to interject material when needed to understand a specific topic in Amos. My experience is that students follow the argument better if the background material is presented first.

As the subtitle indicates, this book takes a socio-historical approach to Amos. In recent years, the emphasis in prophetic studies has been on literary analysis, employing especially form, rhetorical, and redaction criticism. Rather than dealing with the historical prophet, scholarship has concentrated on interpreting the book as a literary work. While not discounting the value of such criticism, recent investigations into the social history of preexilic Israel reveal that a remarkable number of oracles in the Book of Amos fit the period of eighth-century Canaan. Biblical scholars need not be so skeptical about recovering the historical Amos. The view of Amos and his message that emerges from this approach will, I believe, surprise the reader.

It is impossible to mention by name all who have contributed to this study of Amos. I now have a new awareness that scholarship is a communal activity. I am grateful to the many students in my classes at Davidson College who have raised penetrating questions and forced me to rethink the contributions of the Hebrew prophets. I appreciate the criticism of Andrew Dearman, Carl Evans, and Franklin Jacks, who came to my aid in the preparation of this book. My son Vance, a Presbyterian minister at the beginning of his career, offered me valuable assistance by analyzing every paragraph of the manuscript and insisting that I state my position clearly. And I must express my special gratitude to E. Theodore Mullen, Jr., who read and reread the manuscript and offered many helpful suggestions for revision. To Davidson College I am indebted for granting me a sabbatical that made available the time so needed for research and writing. I appreciate the careful work of the editorial staff at Oxford University Press in preparing the manuscript for publication. In a special category, I express my thanks to my wife Jackie, whose patient encouragement supported me over the years it took to complete this book.

Davidson, N.C.
May 1988

M. E. P.

Contents

Abbreviations

BA	*Biblical Archaeologist*
BAR	*Biblical Archaeology Review*
BASOR	*Bulletin of American School of Oriental Research*
BJRL	*Bulletin of John Rylands Library*
BZAW	Beihefte zur *ZAW*
CBQ	*Catholic Biblical Quarterly*
Exp Tim	*Expository Times*
HTR	*Harvard Theological Review*
HUCA	*Hebrew Union College Annual*
IB	*Interpreter's Bible*
IDB	*Interpreter's Dictionary of the Bible*
IEJ	*Israel Exploration Journal*
JAOS	*Journal of the American Oriental Society*
JBL	*Journal of Biblical Literature*
JNES	*Journal of Near Eastern Studies*
JNSL	*Journal of Northwest Semitic Languages*
JQR	*Jewish Quarterly Review*
JSOT	*Journal for the Study of the Old Testament*
JSS	*Journal of Semitic Studies*
JTS	*Journal of Theological Studies*
NEB	New English Bible
Or	*Orientalia*
OS	*Oudtestamentische Studiën*
RSV	Revised Standard Version
SR	*Studies in Religion/Sciences Religieuses*

VT	*Vetus Testamentum*
VTSup	*Vetus Testamentum*, Supplements
ZAW	*Zeitschrift für die alttestamentliche Wissenschaft*
ZTK	*Zeitschrift für die Theologie und Kirche*

Amos and the
Davidic Empire

1

Introduction

SINCE I BEGAN teaching Amos, I have wondered why a prophet from the southern kingdom of Judah went north to condemn Israel for its apostasy. Was the north more wicked than the south? If so, what was the nature of its wickedness? Most books on Amos do not address this issue. When they do, they simply state that either Baal worship was more prevalent in the north than in the south, or the economic prosperity of the north had led to a greater oppression of its poor.

The thesis of this book is that Amos, a prophet of the southern kingdom of Judah, went north to condemn Israel's division of the Davidic kingdom. Amos believed the north had placed illegitimate kings upon its throne by forming a rival monarchy to the Davidic dynasty. These rival kings not only supported unauthorized Yahweh cult centers, but they also ruled their people unjustly. Amos' goal was to persuade the north to rejoin the south, again forming one nation under the rule of a Davidic king. As Yahweh's true representative, the king would focus all worship of God and all administration of the kingdom at the most holy site in Canaan, Jerusalem. Only this could make possible true justice and a lasting prosperity. With the north's rejection of his message, Amos began to proclaim that nation's destruction and exile. However, Amos also believed that eventually the righteous remnant of that nation would join with the southern kingdom to form one nation under the Davidic king, a nation that worshiped Yahweh only in Jerusalem and that received the divine blessings of justice, peace, and prosperity.

This is a somewhat novel approach to the Book of Amos.[1] Because

3

Hebrew faith became universal and monotheistic in the postexilic period, these lofty concepts have commonly been ascribed to the preexilic period. This trend is especially true in prophetic studies. With the prophets at the apex of Hebrew religion, must they not have believed in one God who was sovereign over all nations? Amos is identified frequently as the first exponent of ethical monotheism among the classical prophets.[2] Amos 1:1–2:3 and 9:7 are used to support the view that Amos believed Yahweh, the God of Israel, is also the one true God over all nations.[3]

Yet the primary thrust of this study is with Amos and state religion. These are the historical and biblical roots of many contemporary issues. We know that throughout the Ancient Near East religion was intimately related to the state. Nowhere did kings rule without the approval of the local deities; in turn, the kings supported the local shrines and their priesthoods. Each country developed its own state religion. Israel was no exception, because Yahweh had a special relationship with his people. To worship Yahweh before the exile a person had to be a member of the Hebrew nation. Only after the exile, with the destruction of Israel, was the Hebrew religion divorced from such close ties to the nation-state. Judaism became the name given to the Hebrew religion from the postexilic period to today.

Despite this separation of religion from a geographic identity, Judaism has always had a nationalistic predisposition. Jews of the dispersion celebrated Passover with the words "next year in Jerusalem," expressing their hope for a restoration of the state. With the secularization of Judaism in the Western world, the phrase took on a more symbolic meaning as the universal nature of Judaism was emphasized and its nationalistic tendencies were minimized. This transformation reached its climax in the first quarter of the twentieth century when Reform Judaism defined its faith in totally universalistic terms.[4] However, the double threat of assimilation through secularization and of annihilation through the Holocaust forced Judaism to return to its nationalistic roots.

Before the creation of the present State of Israel in 1948, the term "Israel" in the Jewish prayer books referred to either all Jewish people or the community of the faithful. But since 1948, the term bears still another meaning; it often refers to the State of Israel. Thus modern Judaism is once again reaffirming aspects of its ancient nationalism.[5] No person can understand the depth of present crises in the Middle East without some background knowledge of Jewish nationalism.

A volume of this length could easily be devoted to either the literary structure of the Book of Amos or the prophet Amos within his socio-historical setting. I have chosen to do neither exclusively. The first two sections of this chapter introduce the current key issues in the studies of Amos: the first section outlines the major options for a literary analysis of the Book of Amos; the second section extracts information on Amos himself from the prophetic book that bears his name. The third section gives a summary of each chapter of the book's essential argument.

This study of the Book of Amos employs the primary tools of biblical criticism. At times textual criticism determines the original wording of a passage. While literary and historical criticism help to establish the authenticity of certain passages, form criticism techniques uncover the structure and purpose of a passage within the total message of Amos. Tradition criticism reveals the cultural heritage from which Amos drew in conveying his message. Redaction criticism enhances our respect for the editors of the various versions of the book that have made the words of Amos relevant for other ages. Rhetorical criticism suggests the inner structure of an oracle for more effective oral communication. Historical criticism assists in placing both the prophet Amos and specific oracles within their original settings.[6] Each specific critical method will become clear from the treatment of the material. Technical details are minimal within the text, but scholarly explanations are contained in the notes.

The Book of Amos

"The *words* of Amos, who was among the shepherds of Tekoa, which he *saw* concerning Israel . . ." (Amos 1:1; italics mine).[7] The book's opening verse contains an intriguing anomaly: Amos "sees" his words. This peculiar image forms a prelude to the structure of the entire Book of Amos, because the first six chapters contain prophetic oracles, and the last three recount visions. The oracles are divided between those denouncing foreign nations (1:3–2:3) and those condemning Israel (2:6–6:14). The final chapters contain accounts of five visions (7:1–3, 4–6, 7–9; 8:1–3; 9:1–4).

That Amos wrote any of this material himself is doubtful. It was

probably transmitted orally but was not written until sometime after
722/1 B.C.E.[8] At that time, when Assyria had destroyed northern
Israel, Amos' words of condemnation assumed the status of fulfilled
prophecy. Because the text gives no hint that Amos had followers
either in the north or in the south to preserve his oracles, their oral
"prehistory" remains necessarily clouded. There are indications,
however, that later editors both organized and made additions to the
text. Scholars differ on how much they attribute to these redactors.

In recent years Robert B. Coote in *Amos among the Prophets* (1981)
takes the most radical approach to the composition of the Book of
Amos. Through the use of form, redaction, and historical criticism (especially an examination of sociological history), Coote recovers three
stages in the composition of the Book of Amos. He ascribes Stage A to
the eighth-century prophet Amos. This portion consists of forty verses
that condemn, in scathing language, the ruling elite of Israel for inflict-
ing economic oppression on the poor. No theme of hope is present; the
upper classes are doomed to destruction. Stage B is not an *addition* to
Amos A, but is rather a *new edition* of Amos that "reactualized" the
prophet's message for a new age. The B-stage editor applied eighth-
century prophetic traditions (including some from Amos) to important
issues in Judah during the period from Hezekiah to Jehoiakim. Coote
calls the B-stage writer the Bethel editor. Probably a member of Josiah's court, the B-stage editor accepted the deuteronomistic[9] position
that Jerusalem was now the rallying center for the Mosaic faith. He
agreed with Josiah's reforms, especially the destruction and desecration of Bethel (I Kings 13:1–32; II Kings 23:15–20); but on the basis of
covenant justice, he also opposed the existing ruling class in Jerusalem.
The Bethel editor called for repentance by having Amos condemn Israel's ruling classes with the message "repent or be destroyed." When
the north did not repent and was destroyed, that historical reality
should warn all B-stage readers that the same will occur to them if they
refuse to change. Finally, Coote dates the C stage in the exilic period,
because he regards its content as a midrash on Amos B. This stage offers
comfort and hope to the exilic community. God will return the exiles to
the land not to restore a ruling elite who oppress the poor, but to uplift
the downtrodden and to offer his people the good life of field and
vineyard.

Few scholars agree with the quantity of material Coote attributes
to later editors of the book; however, most scholars acknowledge
later additions to the text.[10] For example, the historical introduction

(1:1) is undoubtedly the work of a later editor.[11] It assumes Jeroboam must be identified as "the son of Joash, king of Israel" (that is, he is *not* Jeroboam I). Contrast this with 7:11, in which he is called simply Jeroboam, as if the one hearing the oracle knew the king could only be Jeroboam II. For stylistic and historical reasons, 1:1; 2:4–5, 10–12; 3:1b, 7; 5:25–26; 6:1a; and 8:11–12 are often attributed to the deuteronomistic editors. Most scholars regard 9:8a as the final verse of the original prophecies of Amos; 9:8b–15 are likely postexilic, hope additions. The denunciation of Edom in 1:11–12 probably comes from a later Judean editor, because it reflects the period of bitter conflict with Edom after the destruction of Jerusalem by the Babylonians in 587. Largely on stylistic grounds, many also judge the oracle against Tyre (1:9–10) a later addition. Moreover, dating the doxologies (4:13; 5:8–9; 9:5–6), which present God as Creator, has stimulated extensive, prolonged debate. On literary grounds, some scholars maintain the doxologies are intrusions into the text; others claim success relating them to surrounding materials. Some scholars argue on theological grounds that any reference to Yahweh as Creator must be given a postexilic date to reflect the influence of Deutero–Isaiah, Job, and possibly the Priestly account of creation (Gen. 1:1–2:4a). Others regard the creation motif as part of Israel's faith since the entrance of the Hebrew people into Canaan.

The styles of Amos' sayings are remarkably varied in form. Most common is the messenger formula: it usually begins with "Thus Yahweh has said" (1:3, 6, 9, 11, 13; 2:1, 6); it may have an internal or concluding formula "oracle of Yahweh" (2:11, 16; 3:10, 13); it usually closes with "Yahweh has said" (1:5, 8, 15; 2:3, 16). Another style is the proclamation formula "Hear this word" (3:1; 4:1; 5:1). It calls listeners to hear an announcement; it is especially effective when an indictment follows. The oath formula "Yahweh has sworn by" is always a prelude to words of judgment (4:2; 6:8; 8:7). The mood is critical when disputation sayings are cast in the form of questions, with obvious answers (3:3–6; 9:7). A funeral dirge (5:2) or a series of woe saying (5:18; 6:1 plus 3–6; perhaps 5:7) strike somber moods.

Either Amos himself or, more likely, an editor has organized the oracles and visions into effective units. For example, punctuating the oracles in 1:3–2:8 is the recurring formula "For three crimes and for four, I will not cause it to return" (author's translation). A repetition of the ominous words "yet you did not return to me" follows each successive curse sent by Yahweh in oracles 4:6–12. Four of the five visions, furthermore, are arranged in pairs of two: the first pair contains the hopeful words "Yahweh repented concerning this" (7:1–6), whereas the second pair contains the hopeless words "I will never again pass by them" (7:7–9; 8:1–3).

Whereas most scholars would regard Amos as neither a professional cult figure nor a court sage, the style of some authentic Amos' sayings do reflect cultic and wisdom influences. A hymnic style similar to the Psalter occurs in 1:2; 4:13; 5:8–9; 9:5–6. The use of graduated numbers ("For three crimes and for four") is a hallmark of the wisdom tradition (see Prov. 30:15–16, 21–23, 29–31 for striking parallels). In addition, three other sets of verses suggest the didactic style popular in clan wisdom: the long series of questions in 3:3–6 that use simple analogies drawn from peasant life to convey God's wrath against Israel, the short series of questions in 6:12 that reveal the perverse nature of sin, and the admonitions in 5:14–15 that call for high ethical conduct. The evidence demonstrates that Amos was no illiterate peasant, because he made skillful use of the rhetorical styles current in his day.

The Prophet Amos

The deuteronomistic historians and the chronicler narrated many stories about the prewriting prophets. Their histories preserve legendary tales about Gad, Nathan, Ahijah, and especially Elijah and Elisha. We are told more about the lives of these prophets than about what they proclaimed.

With Amos, the first of the classical prophets, the situation is reversed. Not only does no reference to Amos exist outside the biblical book bearing his name, but that book itself records primarily the prophet's words rather than his deeds. The deuteronomistic historians did know about Amos; evidence points to their editing an early version of the Book of Amos. In their history of Israel, the deuteronomists simply chose not to mention Amos.[12] We are, therefore, totally dependent on the Book of Amos for information about the prophet.

According to the deuteronomistic introduction (1:1), Amos prophesied during the reigns of Uzziah of Judah (ca. 783–742) and Jeroboam of Israel (ca. 786–746), son of Joash. Most scholars accept this as historically accurate, dating Amos ca. 760. Amos' message fits the period of prosperity under Jeroboam II. Furthermore, the extent of the kingdom of Jeroboam II (II Kings 14:25) is identical with that given in Amos 6:14b.

Some scholars argue for a possible date during the reign of Tiglath-pileser III (ca. 745–727), who campaigned in Syria and captured Damascus in 732. The evidence cited are references in Amos to destruction coming from the north and the exile of surrounding populations (1:5; 4:2; 5:5; 7:11, 17). These scholars maintain Tiglath-pileser III was the first to implement mass exiles; previously, only the rulers of defeated nations were exiled.[13] But this overlooks evidence that Urartu exiled large populations during an earlier period.[14]

Numerous references in the Book of Amos to an earthquake (1:1; 3:15; 4:13; 6:11; 8:8; 9:1, 5) support a date ca. 760. Yigael Yadin, through his excavations of Hazor (Stratum VI), dates an earthquake to the first half of the eighth century.[15] This archaeological evidence coincides with the period designated in the deuteronomistic introduction.

Amos' preaching at Bethel and Samaria has led some scholars to place the prophet's origin in Israel, but this theory has won little support.[16] Most scholars believe he was from Judah, probably a resident of Tekoa, a small village six miles southeast of Bethlehem and ten miles south of Jerusalem. Because the text reads that he was "among the shepherds of Tekoa" (1:1), he may have been a native of some other village who merely joined the Tekoa shepherds to make his living. The questioning of Amos' origins is probably too skeptical. Little internal evidence in the Book of Amos questions the prophet's connections to Tekoa. Also, because a passage from II Samuel demonstrates a strong wisdom tradition in that village, one may conclude Tekoa was his home. II Samuel records an interesting account of David's general Joab who sent for a wise woman from Tekoa to heal the breach between David and his son, Absalom (II Sam. 13:37–14:24). The passage is laced with references to wisdom, not least of which are the woman's words in defense of Joab's actions.

> We must all die, we are like water spilt on the ground, which cannot be gathered up again; but God will not take away the life of him who devises means not to keep his banished one an outcast.
>
> (II Sam. 14:14)

Perhaps in the village of Tekoa Amos encountered the clan wisdom found in his oracles.

When Amos is called a "shepherd," one should not simply assume he was a peasant, a lowly keeper of sheep in the wilderness area surrounding Tekoa. The Hebrew word for shepherd *(nōqēd)* only occurs one other place in Hebrew Scriptures. In II Kings 3:4 the word is used to describe Mesha, the king of Moab, who had to deliver annually to the king of Israel 100,000 lambs and the wool of 100,000 rams! Although the accuracy of this claim cannot be determined, the term *nōqēd* leaves open the possibility that Amos was a wealthy sheepbreeder rather than a lowly peasant herdsman. The biographical section of the Book of Amos records the prophet's reference to himself as a herdsman (7:14; *bôqēr*). The Hebrew word *bôqēr* means "one who herds cattle," indicating that Amos may have grazed large cattle as well as small sheep. In written Hebrew *bôqēr* and *nōqēd* are similar. In fact, *bôqēr* may simply be a corruption of *nōqēd*, because in 7:15 Amos said that God took him from following the flock, a term associated with sheep.

The Ugaritic texts use the term *nqdm* for officials who have something to do with sheep. These officials have a higher social standing than ordinary shepherds. While some were associated with the temple, the majority were servants of the royal establishment. Possibly Amos cared for a flock of sheep that were owned by the temple in Jerusalem.[17] As a member of the temple personnel, he would have had contact with cultic language similar to that of his sayings.[18] But more likely he was a royal sheep manager who engaged in business transactions in Israel's northern marketplaces.[19]

Amos was also a "dresser of sycamore trees" (7:14). Although the fruit of the sycamore tree was not considered as fine as figs, it could be made sweeter by puncturing it before it ripened. The fruit then fermented and became more edible. Sycamore trees do not grow in the hill country around Tekoa, but are common on the coastal plains. Amos may have owned a grove of sycamore trees. More likely he was hired to puncture the fruit of another person's sycamore trees. Whether Amos was a shepherd or a "dresser of sycamore trees," his true economic status is impossible to determine. That his sympathies lay with the poor and the disinherited of the land does not prove he was from that poorer class.

The Book of Amos contains numerous references to the prophet as visionary.[20] Amos 1:1 refers to words "which he saw *(ḥāzâ)* concerning Israel." Thus, the prophet's visions carried with them a prophetic word that Amos was commissioned to deliver. In the conflict with Amaziah at Bethel (7:10–17), the northern priest referred to

Amos as a seer (*ḥōzeh*). Given the context, Amaziah undoubtedly believed Amos was a member of a professional guild of seers known as *ben-nĕbî'îm*, the "sons of the prophets." These prophets were paid to have visions and offer counsel, as in I Samuel 9:5–24 where Samuel functions as a seer to help Saul find his father's lost asses. Visions are a part of Amos' prophecy as demonstrated in chapters 7–9, which contain a series of five. Each vision (locusts, 7:1–3; fire, 7:4–6; plumb line, 7:7–9; basket of summer fruit, 8:1–3; altar, 9:1–4) conveys God's wrath against Israel.

The mythic context for the vision is probably the Council of Yahweh, a heavenly assembly of spiritual beings. The council surrounds God and enhances his glory; it provides a court befitting his majesty. Members of the council advise God concerning his best course of action and do his bidding.[21] The terminology used in Scripture to identify both the council and its membership is strikingly similar to that found in Mesopotamia and Canaan for the assembly of the gods.[22] The Hebrew people apparently borrowed the concept from their neighbors, but demoted the members of the divine assembly from the status of independent spiritual beings. In this way they resisted the temptation to allow council members to rival God. God speaks to these spiritual beings in Genesis 1–11, in which he uses the divine plural: "Let us make man in our image" (Gen. 1:26); "Behold the man has become like one of us" (Gen. 3:22); "Come, let us go down, and there confuse their language" (Gen. 11:7).

The vision of Micaiah (I Kings 22:19–23) and the calls of Isaiah of Jerusalem (Isa. 6), Deutero–Isaiah (Isa. 40:1–11), and Ezekiel (Ezek. 1–3) are the best examples in Scripture of the prophets within the council meeting. They do not become members of the council; rather, they are called into God's presence, receive the vision, and are commissioned to prophesy.[23] Whereas the Book of Amos contains no reference to the council,[24] that body provides the appropriate context for the visions.

Two passages recount the call of Amos into the prophetic ministry. In the biographical unit (7:10–17), the confrontation with Amaziah at Bethel provided the occasion for Amos to defend his preaching. He did so by offering his "prophetic credentials."

I am no prophet, nor a prophet's son; but I am a herdsman, and a dresser of sycamore trees, and the LORD took me from following the flock, and the LORD said to me, "Go, prophesy to my people Israel."[25]

(7:14–15)

Amos' reply is subject to various interpretations. Verse 14 is a verbless sentence in Hebrew. The tense must be provided from the context. Was Amos denying he was a prophet? "I am no prophet, nor a prophet's son." Was he stating that, although he was not originally a member of a prophetic guild, nevertheless, God called him? "I was no prophet, nor a prophet's son . . . but the Lord took me from following the flock, and the Lord said to me, 'Go, prophesy to my people Israel.' " John D. W. Watts' interpretation in *Vision and Prophecy in Amos* stresses the importance of the words "took me" and "go, prophesy." Watts believes 7:14–15 should be translated:

> No prophet did *I* choose to be! (I did not choose to seek the status of *nābî*.) Nor did *I* seek to become one of the prophetic guild. For *I* (had chosen to be) a herdsman and a tender of sycamores, when *Jahweh* took me from following the flock (the place of my choice). But it is *Jahweh* who said to me, Go! Be a prophet to my people Israel!²⁶

Because Amos had proclaimed at the northern king's sanctuary that Jeroboam II will be slain and that Israel will be exiled from the land, Amaziah charged him with conspiracy to overthrow the government. The modern reader has difficulty understanding the charge. But to the Hebrew mind, the proclamation of the word of God by a true prophet released into history God's actions. God's word had power to create that which it signified.²⁷ The Hebrew term for "word" (*dābār*) can also be translated "event," "thing," or "happening." A word of doom by a true prophet in the form of a curse or judgment oracle began the process of releasing God's wrath in history.²⁸ Amos' announcement of the coming destruction of Israel was taken seriously by Amaziah. Amos had threatened the foundation of the northern kingdom with his attack on the temple at Bethel that supported the northern monarchy.²⁹ Amaziah's loyalty to his ruler demanded that he denounce Amos and refuse to let him speak again at Bethel. After all, was it not a seer, a disciple of Elisha, who anointed Jehu and caused the bloody overthrow of the Omride dynasty (II Kings 9:1–13)? Was not the present ruler of Israel a member of the Jehu dynasty? Given the traditional conflict between northern and southern kingdoms, Amaziah assumed Amos had been sent north as an official prophet from the Jerusalem court to announce the destruction of

Israel. Amaziah demanded that Amos return to the south to prophesy.

The second reference to Amos' calling is the oracle in 3:3–6, 8.[30] The unit forms a didactic disputation consisting of a series of nine questions arranged in climactic order. The first five questions call for negative answers. Then the style abruptly changes with the last four questions, the final two of which refer to Amos' prophetic calling. Beginning with 3:3, we have a preface with no reference to doom:

> Do two walk together,
>   unless they have made an appointment?

The next verse initiates a growing sense of doom as one animal overpowers another:

> Does a lion roar in the forest,
>   when he has no prey?
> Does a young lion cry out from his den,
>   if he has taken nothing?

In the following verse a human hunter catches an animal:

> Does a bird fall in a snare on the earth,
>   when there is no trap for it?
> Does a snare spring up from the ground,
>   when it has taken nothing?

In 3:6a the theme of doom intensifies with a reference to the destruction of a city:

> Is a trumpet blown in a city,
>   and the people are not afraid?

Verse 3:6b contains the first reference to God's wrath:

> Does evil befall a city,
>   unless the LORD has done it?

Verse 3:8 follows naturally upon 3:6. It is Amos who is called to declare what God intends to do:

The lion has roared;[31]
who will not fear?
The LORD God has spoken;
who can but prophesy?

Synopsis of Argument

Chapters 2 and 3 describe the place of kingship in state religion within the Ancient Near East. An examination of Egyptian, Mesopotamian, and Canaanite cults reveals that the gods were viewed as establishing kingship. Three major functions of kings were to build temples honoring the god(s) who placed them on the throne, to secure national freedom by defeating all foreign enemies, to establish justice in the courts. When these functions were satisfactorily accomplished, the nation prospered. Kingship in Judah and Israel fits this pattern; the ideal king will protect his people from their enemies, establish justice in the land, and honor Yahweh through the maintenance of the cult. The Davidic–Solomonic kingship introduced state religion to Israel. When the kingdom was divided, northern Israel developed its own state religion. It established a rival dynasty (Jeroboam I), a rival capital (first Shechem, then Tirzah, and finally Samaria), two rival sanctuaries (Bethel and Dan), a separate priesthood, and a new festival calendar. Two conflicting state religions emerged, both dedicated to worshiping the same God.

Chapter 4 explores Amos' attitude toward neighboring nations. The chapter begins with an examination of whether all the oracles against foreign nations should be attributed to Amos and proceeds to analyze the form of these oracles. Next comes a consideration of five theories that have been proposed to explain Amos' concern over the crimes committed by foreign nations:

1. Amos was a monotheist.
2. Amos was a universalist.
3. Amos employed the standards of an international law.
4. Amos employed the style of Egyptian Execration Texts.
5. Amos supported a reunited Davidic empire.

The last theory offers the best context for interpreting Amos' oracles against foreign nations. Each oracle is then interpreted within the setting of a disintegrated Davidic empire.

Amos' condemnation of worship at northern shrines is the subject of Chapter 5. Four major positions concern Amos' attitude toward northern religion:

1. Amos opposed all cultic worship.
2. Amos opposed the substitution of sacrifices for social justice.
3. Amos criticized Canaanite influences in the northern cult.
4. Amos rejected northern sites as legitimate places of worship.

While only the first position is found unacceptable, the fourth position best explains Amos' attack on the northern cult centers. After examining those passages related to Amos' attack on worship at Bethel, Gilgal, Samaria, Dan, and Beer-sheba, we conclude that Amos called Israel to seek God in Jerusalem.

Chapter 6 explores the basis for Amos' condemnation of the upper classes in Israel for exploiting the poor and the needy in the land. Recent scholarship on the judicial systems in both Israel and Judah enhances our understanding of Amos' call for social justice and righteousness. Especially relevant are the judicial reforms of Jehoshaphat, which probably influenced the court system in Tekoa. Amos believed that God had entrusted to the Davidic line the responsibility for fostering justice in a monarchical judiciary system. But the northern rebellion had resulted in the establishment of kings lacking legitimacy who ruled Israel unjustly. Amos condemned this northern judicial system. He believed the only way to establish a truly just court system was through acceptance of a united kingdom ruled by the Davidic kings.

Did Amos proclaim the inescapable destruction of Israel or did he offer that nation one last chance to repent and be saved? This is the subject of Chapter 7. An analysis of three oracles and two visions show that in an early stage in Amos' ministry he interceded for Israel and urged God not to destroy it. During this period, Amos believed that if that nation would worship God in Jerusalem, accept the Davidic king as its rightful ruler, and establish justice and righteousness in the courts, then God would bless it.

The subject of the final chapter is Amos' view of Israel's destiny, the view of the Day of Yahweh. We consider three theories for interpreting the Day of Yahweh: Mowinckel's cultic approach, von Rad's holy war position, and Frank Cross' divine warrior motif. The mythico-historical theory of Cross provides the most productive approach for understanding the Day of Yahweh in Amos: That day will result in the total destruction and exile of Israel. However, beyond destruction Amos saw a new Day of Yahweh fulfilled in a reconstituted Davidic empire. In that reborn nation, the righteous remnant of Israel will reunite with Judah, worship God in Jerusalem, and assure that peace, prosperity, and justice are achieved through accepting the legitimate rule of the Davidic dynasty.

2

Kingship and State Religion in Egypt, Mesopotamia, and Canaan

ONE OF THE few events recorded from the life of Amos is his conflict with Amaziah at Bethel (7:10–17). Amaziah, a priest of Israel, presided over a royal sanctuary established in support of the northern monarchy. Monarchy and state religion are intrinsically related in the text. Amos' attack against the northern dynasty (v. 16) and its priesthood (v. 17) demonstrates the prophet's belief that both institutions were contributing to the downfall of the northern kingdom. Amaziah regarded the threat sufficient to notify King Jeroboam II of Israel that a conspiracy was taking place and to command Amos to return to Judah where he could conduct his prophetic profession without opposition.

The conflict between Amos and Amaziah accentuates the relationship between kingship and state religion in both Israel and Judah. Before examining this subject in Chapter 3, we must first consider the wider context of state religions in the Ancient Near East. Our focus in both chapters will be on three major functions of kingship in the Ancient Orient: the dedication of temples and the maintenance of cults to secure divine blessings upon the land, the conquest of foreign foes to assure the nation's freedom, and the administration of justice in the courts to establish law and order among the people. All three functions reveal a close relationship between kingship and the gods. Nowhere did kings rule without the approval of local deities; the kings' duties were to govern in a manner that honored the gods who had placed them on their thrones. Kings and gods together assured the well-being of nations.

17

Kingship and State Religion in Egypt

In Egypt the pharaoh was considered divine, the son of various gods. The meaning of "divine" is vague. Formed by the gods in the queen's womb, some scholars attest to his divinity before birth, a god by nature.[1] The pharaoh was the embodiment of Horus, the falcon god; he was the son of Re, the sun god; in the imperial age, he was the son of Amon, now identified with Re.[2] After death the pharaoh was equated with the god Osiris, and his successor with the god Horus. A cartouche of Senusert III states his divine nature through his official titles: "Horus, 'Divine of Forms'; Two Ladies, 'Divine of Births'; Horus of Gold, 'Who becomes'; King of Upper and Lower Egypt, 'the Ka's of Re appear in Glory'; Son of Re, Senusert; granted life and wealth eternally."[3] In the Amarna Letters, the Canaanite vassals address the pharaoh as "my Sun-god," before whom they prostrate themselves "seven times and seven times."[4]

Other scholars contend that too much emphasis has been placed on the pharaoh's relationship to the gods.[5] Although the pharaoh was regarded as divine, he was no more than a demigod because his mother was human. He was the reflection in this world of the divine principle, but he was not a god incarnate. He was not above divine law, but was bound by it as were all human beings. At times the people resisted the pharaoh, even questioning his right to rule. The precise nature of the pharaoh's divinity remains enigmatic.

The pharaoh was high priest of the land, although he delegated much of his cultic authority to priests who conducted services on his behalf. Only at major religious festivals and in the consecration ceremonies for temples he had constructed did he personally participate.[6] Routine ceremonies were performed by other priests.

He was commander-in-chief of the army. Although not all pharaohs engaged in military campaigns, they are pictured in Egyptian art as mighty warriors. A typical pose depicts the pharaoh holding a group of captives by their hair in order to slay them or riding a chariot over a battlefield of fallen enemy troops.[7]

He also served as chief justice of the nation, responsible for the proper administration of the courts. No law codes have been discovered in Egypt; the king was the absolute lawgiver who ruled by personal divine decrees. He usually entrusted the implementation of these decrees to local officials.[8] The Egyptian word for "justice," "order," "truth" is *ma'at*, which in its most inclusive sense is related to the whole ordered structure of creation. In one creation myth, the powers of chaos threaten the order of the universe before they are subdued by Re. Thus Re was known as the "Lord of *ma'at*." As the divine representative of Re, the pharaoh was expected to establish justice and order on the earth.[9] His titles—"the good shepherd," "a roof," "a mountain," "a fortress," and "a refuge to the homeless"— reflect this responsibility. An enthronement hymn for Merneptah reads:

> Rejoice, thou entire land: the goodly
>   time hath come.
> The Lord has appeared in the Two Lands.
>   * * *
> Merneptah, contented with Truth.
> All the righteous (say): Come and behold:
> Truth hath repressed falsehood,
> the sinners are fallen on their faces,
> all that are covetous are turned back.
> The water standeth, and faileth not, the Nile
>   is running high.
> The days are long, the nights have hours, the
>   months come aright.
> The gods, are content and happy of heart,
> and life is spent in laughter and wonder.[10]

The pharaoh appointed a vizier whose chief function was to administer justice. This official's duty was not only to hear cases in the judgment hall, but also to go out among the poor and provide them opportunities to present their cases. Local officials conducted courts of justice under the vizier's oversight.[11]

Kingship and State Religion in Mesopotamia

The view of the king's relationship to the gods in Babylonia and Assyria was somewhat different from that in Egypt. Deity was not ordinarily ascribed to the king; yet kingship was regarded as an essential institution in both human and cosmic affairs. Kingship was considered an ancient gift of the gods; it came to earth from heaven.[12] The people used highly exalted language to refer to their kings. A number of kings have the eight-pointed star (the divine determinative) prefixed to their names[13]; their divinity, however, was by *adop-*

tion rather than by *nature*.[14] The kings assumed the role of divine representative at their coronation. This divinity was functional rather than metaphysical.[15] Mesopotamian kings certainly regarded themselves as sons of gods. Hammurabi, in the preamble of his code and in other texts, referred to himself as "son of Sin," "son of Dagan," and "son of Marduk." But these titles probably convey the king's dependency on the gods and the king's godlike qualities rather than his actual deification.

The king's role in the state religion was especially prominent in the recitation and reenactment of the Babylonian Creation Epic (*Enuma Elish*).[16] Though popularly known as the Creation Epic, it is more accurately entitled "The Exaltation of Marduk," Babylon's chief god. The stage is set for Marduk's ascendancy when Tiamat (primordial waters) and her consort Kingu, along with an army of monsters, war against the great gods of the orderly cosmos. Tension mounts when the ancient gods Ea (earth) and Anu (sky) are forced to retire. The situation grows desperate when Tiamat seizes the tablets of fate, which determine the unchangeable state of things for humankind, and places them in the hands of her consort. At an emergency session, the assembly of the gods elects the young and virile Marduk as king. Marduk delivers the gods from chaos, first by slaying Tiamat and Kingu and second by recovering the tablets of fate. From the body of Tiamat, Marduk fashions the heavens and the earth; he counsels Ea to create humankind from the blood of the rebel god Kingu. The epic climaxes with the building of Ersagila, Marduk's temple in Babylon, where the decreeing of the fates by Marduk[17] and the recitation of Marduk's fifty glorious names[18] exalt him among the gods.[19]

The Babylonian New Year Festival (*Akitu* Festival) was an eleven-day reenactment of this creation epic. The emphasis was entirely on renewal, which befits a New Year celebration. The king's role in the drama is ambiguous. In an earlier period he possibly functioned as a priest–king and assumed the role of Marduk, defeating the forces of disorder and helping the gods fix the cosmic order for the coming year.[20] In later times the priest undertook this role, while the king served as a master of ceremonies and recited parts of the Creation Epic.

At the beginning of the festival, an official took the hand of Marduk's statue and led it in procession through the Ishtar Gate and across the Euphrates River where the celebration took place. At the close of the festival, the same statue was returned to Babylon and reinstated in the temple. As in the epic, the tablets of fate were first seized by Tiamat and given to Kingu, but later reclaimed. The temple Ersagila, which had been momentarily abandoned by Marduk, was once again rededicated to him when he returned triumphant.[21] Renewal of life was then celebrated through the ritual marriage of Marduk in the temple.

On the New Year Festival's fifth day, kingship was renewed through the ritual of reenthronement. The king played a prominent role.[22] The high priest divested the ruler of the scepter, the circle, and the sword (all symbols of royal office) and struck the king's cheek as if he were a commoner. The king, dragged by the ear before the statue of Marduk and forced to kneel, at once proclaimed his loyalty to Marduk. The priest then restored the scepter, the circle, and the sword to the king and struck the royal cheek with enough force to bring tears, a good omen for the coming new year.[23]

We should not assume that the Babylonians believed this cultic ritual controlled the gods; rather, the temple rites made concrete what was taking place in the mythical realm.[24] The king became sovereign on earth because Marduk was sovereign in heaven. The New Year Festival was therefore an essential part of state religion in Mesopotamia.

The king's role as a victorious warrior over the nation's foes is dramatically conveyed in "The Forest Journey" from the *Epic of Gilgamesh*. In this adventure Gilgamesh traveled to the land of cedars (Lebanon?) and destroyed the evil monster Humbaba. The monster represents not only the chaotic forces of evil, but also the nation's enemies. As the legendary king of Uruk, Gilgamesh, a demigod who was two-thirds god and one-third man, was empowered to win "victory in battle from which there is no going back."[25]

As Marduk's representatives on earth, the kings of Babylon had authority to govern their people. They codified the laws of the people, attributed them to the gods, and theoretically lived under these laws.[26] As in Egypt, the king's responsibility was to maintain law and order. An ancient Sumerian hymn (ca. 1916–1896 B.C.E.) glorifies the wisdom and righteousness of a local ruler:

> Make the good day dawn like Utu,
> all eyes are directed towards you,

all foreign lands will be peaceful under
    your broad protection.

You have made roads (and) paths straight,
filled the land of Sumer with joy,
put righteousness into every mouth,
(and) made the worship of the gods shine forth.

Under your rule men will increase and extend,
hostile lands will rest in peace,
men will enjoy days of abundance.[27]

In the *Epic of Gilgamesh*, Enlil, father of the gods, decrees that
Gilgamesh's destiny is not to become immortal, but to establish justice in the land.

The father of the gods has given you kingship, such is your destiny,
everlasting life is not your destiny. Because of this do not be sad at
heart, do not be grieved or oppressed. He has given you power to
bind and to loose, to be the darkness and the light of mankind. He
has given you unexampled supremacy over the people, victory in battle
from which no fugitive returns, in forays and assaults from which there
is no going back. But do not abuse this power, deal justly with your
servants in the palace, deal justly before Shamash.[28]

Nowhere, however, is the relationship between the gods and the
king's justice better stated than in the prologue to the Lawcode of
Lipit-Ishtar:

When Anu (and) Enlil had called Lipit-Ishtar . . . to the princeship of
the land in order to establish justice in the land, to banish complaints,
to turn back enmity and rebellion by the force of arms, (and) to bring
well-being to the Sumerians and Akkadians, then I, Lipit-
Ishtar, . . . [estab-]lished [jus]tice in [Su]mer and Akkad in accor-
dance with the word of Enlil.[29]

In Mesopotamia, victory in battle, justice, and prosperity depended on the king's relationship to the gods. As the adopted "son
of god," the king defeated his nation's enemies and established law
and order through divine support. Honoring the gods who had placed
him on the throne assured divine benevolence toward the land. "In
principle, the king enables the divine right and justice to be estab-

lished in his kingdom."[30] Mowinckel summarizes this role of the king in Mesopotamia:

> The king is not a god, as in Egypt; but he has been endowed with a divine vocation and with a superhuman power and quality, which in some respects puts him on the same plane as the gods. He possesses a "divinity" shared by no other mortals. He is "like a god"; he is "the image of the gods." He has been filled with divine power, and has authority on earth from the gods, as long as he acts in accordance with their will, with "justice." Hymns may, indeed, be sung in his honour on his festival day; but men do not pray to the king; on the contrary, they pray the gods to bless him.[31]

Kingship and State Religion in Canaan

Canaanite state religion has more in common with Mesopotamian civilization than with Egyptian. Although there is little direct knowledge of the king's function within the cult, there is ample evidence that the king played a major role in establishing justice in the community and assuring prosperity to the land.

The main sources of information about Canaanite culture are the Ugaritic and Alalakh Tablets. The Ugaritic texts, first discovered at Ras Shamra in 1929 and dated in their redacted form ca. 1400 B.C.E., contain mythological tales of the gods (El and Asherah, Baal and Anath, Yamm and Mot) and legendary stories of kings (the Dan'el-Aqhat and the Keret texts).[32] Discovered during the excavations of a site on the Orontes River in 1937–1939 and later in 1946–1949, the Alalakh Tablets contain texts primarily presenting the social order of a Canaanite society in the seventeenth and fifteenth centuries B.C.E.

Two extant legends disclose the Canaanite view of kingship: the Dan'el-Aqhat and the Keret stories. Although both legends present similar views of kingship (kings as demigods), these tales still offer insight into the nature of Canaanite monarchy.

The legend of Dan'el-Aqhat demonstrates the close relationship between the virility of the king and the fertility of the land. The childless King Dan'el petitions to both El and Baal, the two chief gods of the Canaanite pantheon, to grant him a son. Aqhat is conceived and born. When the divine craftsman fashions a magical bow for Aqhat, the gift arouses the jealousy of Baal's consort Anath. She offers Aqhat immortality in exchange for the bow. Aqhat refuses. In anger, Anath transforms the god Yatpan into an eagle and sends him to slay and devour Aqhat. King Dan'el learns of Aqhat's death while holding a court of justice for his people. His curse brings a seven-year drought upon the land. Baal sends the eagle to Dan'el, who opens it, recovers the body of his son, and buries him with mourning rites. Then Aqhat's sister Pegat seizes a sword and dagger to avenge her dead brother. She does meet Yatpan, but here the text breaks off. Aqhat is presumably resurrected, Dan'el ceases mourning for his son, the curse is lifted, and fertility returns to the land.

This cycle of death, sterility, resurrection, and fertility is reminiscent of the Baal–Anath story cycle. Each year at the approach of the dry season in late spring, Baal,[33] the god of storm and rain clouds, is defeated by Mot, the god of death and sterility. Baal then descends into the earth, taking with him the clouds and the rain. Mot sends a somber message to Baal:

> I shall pound thee, consume and eat thee.
> Cleft, forspent and exhausted
> Lo thou art gone down
> into the throat of the god Mot,
> Into the gullet of the Hero, Beloved of El.[34]

The dry summer heat, signifying Mot's victory, comes over the land. The Canaanite pantheon's high god El mourns Baal's death, while Anath wanders in search of him. When Anath finds Baal's body, she performs mourning rites and carries him back to Mount Saphon, the Mount Olympus of Canaanite religion. There Baal is buried and mourned at a great funeral feast.

Baal's throne is occupied temporarily by Athtar (the bright Venus star), son of El's consort Asherah. But Athtar proves a poor substitute for Baal. In rage Anath attacks Mot:

> She seizes the Godly Mot—
> With sword she doth cleave him.
> With fan she doth winnow him—
> With fire she doth burn him.
> With hand-mill she grinds him—
> In the field she doth sow him.
> Birds eat his remnants,
> Consuming his portions,
> Flitting from remnant to remnant.[35]

This passage describes a harvest ritual where new grain is desacralized and made available for common use. The dry season, the season when crops ripen, is over. The harvest is held.

With the change of season, Baal returns to displace Athtar, defeats Mot in a terrible battle, and resumes his throne on Mount Saphon. At this point in the myth Baal and Anath may enter the temple to celebrate the sacred marriage (*hieros gamos*) that is closely associated with fertility worship.

The Dan'el-Aqhat legend and the Baal–Anath myth are strikingly similar. Both the deaths of Aqhat and Baal bring drought upon the land. Just as Dan'el laments the death of his son Aqhat, so do El and Anath mourn Baal's death. The rains return to nourish the earth only after the resurrections of Aqhat and Baal. Both stories relate nature's fertility to the roles of the king and the gods. Given the affinity between myth and reality in the Ancient Near East, the relationship of the king to the productivity of the land is undoubtedly based on his kinship to the gods, the source of all fertility.

The Keret texts emphasize the king's role as victorious warrior, just ruler of his people, and dispenser of fertility. When King Keret loses his palace, his wife, and his children, he enters a state of mourning. Keret offers sacrifices to El, who instructs him to invade the country of Udum and force Huriya, King Pabel's daughter, to become his wife. Next comes a stirring account of the Canaanite king's military expedition. With an army of men from every house and clan, the army camps outside the fortress of King Pabel who, sleepless and terror-stricken, offers Keret royal gifts if he will withdraw in peace. Keret demands and receives the princess Huriya as his queen while Udum's people lament. Sons and daughters are born of this union blessed by El. Years later when Keret grows ill, his son Elhau wonders if his father can rule despite his sickness. Elhau's sister laments the illness of Keret:

> In thy life, our [fa]ther, we rejoiced,
> Exulted in thy not dying.
> (But) like a dog thine aspect is changed,
> Like a cur thy joyous countenance,
> Wilt thou die, then, father, like the mortals,
> Or thy joy change to weeping,
> To a woman's dirge, O father, my song?
> Shall, then, a god die,
> An offspring of the Kindly One not live?[36]

Although the king is the "son of El" ("offspring of the Kindly One"), his severe sickness brings sterility to the land. Nor in his illness is Keret able to administer justice in the nation. Yassib, another son of Keret, urges his father to resign his throne:

> Hear, and may thine ear be alert!
> By slow degrees thou art growing old,
> And in the tomb thou wilt abide.
> Thou hast let thy hands fall into error,
> Thou dost not judge the case of the widow.
> Nor decide the secret of the oppressed.
> Sickness is as thy bedfellow,
> Disease as thy concubine.
> Descend from thy rule that I may be king,
> From thy government that I may be enthroned.[37]

But Yassib has waited too long. El has already restored health to Keret; fertility returns to the land; Keret once again establishes justice in the country. Keret retaliates by calling on the god Horon to curse his rebellious son Yassib.

The emphasis on the king's role in establishing justice in the courts is grounded in El mythology. One of El's major roles is progenitor of the Canaanite pantheon. With his chief consort Asherah, "Creatress of the gods," El became "Father of the gods." In patriarchal fashion, he dwells in a tent, not a palace, and periodically assembles his divine offspring to hear his decrees. El rules by judgments that are executed by members of the divine council. Apparently his decrees are not feared, for El alone among the Canaanite gods reputedly rules wisely; he is called "The Kindly One, The Merciful."

El's tent is on the cosmic mountain from which flows the two rivers, sources of the cosmic waters.[38] It is designated Mount Ll (*not* "Mountain of El"), and may correspond to the majestic Khirbet Afqa of the Lebanon range that has two fast-flowing springs on opposite slopes of the mountain.[39]

A composite of various iconographic sources represents El as a long-bearded male god who sits on a throne. He wears a conical hat with two bull horns; above his head are winged sun disks. His right hand is raised to bless those who honor him. In some reliefs, his throne is in the form of a cherubim (winged sphinxes); in others, he sits upon a plain throne. In patriarchal fashion, he rules the divine assembly wisely and issues decrees.

These two legends of Dan'el-Aqhat and Keret reveal the same three characteristics of kingship we have seen in both Egyptian and Mesopotamian cultures: the king's close relationship to the gods, his victories in battle over foreign nations, and his obligation to establish justice in the courts. That kingship is intimately tied to fertility reflects a central aspect of Canaanite religion. The king is designated the "son of El," a title he assumes when he takes the throne. He establishes justice in the courts and is permitted to reign as long as he remains sexually active. When his wife dies, he can no longer function as king. He exhibits warlike qualities in his successful military campaign to procure a new wife. Once the Canaanite king becomes old and impotent, sterility comes over the land and there is agitation for a new, younger king. This fertility aspect of kingship relates closely to the Baal–Anath myth. The king probably played a prominent role in a ritual celebration of the myth; although, as in Babylon, his place in later times may have been taken by a priest. In any case, the king rules by divine approval and is expected to assure that the nation is free of foreign domination and that justice prevails in the courts.

Canaanite iconography complements the textual sources. An El-stela from Ugarit pictures the god sitting upon the throne and holding audience. Ahiram's sarcophagus from Byblos shows the king or a god seated upon his cherubim throne and ruling over his people. An ivory plaque from Megiddo (Late Bronze Age) depicts a prince upon a cherubim throne surrounded by his attendants.[40] This free pictoral interchange between king and deity confirms those texts that describe the king as the gods' representative on earth.

We have considered how state religion in Egypt, Mesopotamia, and Canaan was rooted in the relationship between the gods and kingship. The responsibilities of the kings were to honor the gods who had placed them on the throne, to deliver militarily the people from their enemies, and to establish justice in the courts. We shall examine in Chapter 3 these functions of kingship as they emerge in the state religion of Judah and Israel.

3

Kingship and State Religion in Judah and Israel

THE EMERGENCE of state religion in Israel is contemporaneous with the establishment of the monarchy.[1] David and Solomon developed a nation-state united around a monarchy supported by the Yahweh cult.[2] The biblical sources for this period are fairly reliable historically.[3] While they are not as factually accurate as modern historians desire, they are historically more dependable than those sources that preserve the mythical and legendary traditions of creation, the patriarchal period, the exodus events, and the conquest and settlement of Canaan.

What emerged in the united kingdom of David and Solomon was a well-developed national religion. There were a divinely appointed monarchy (the Davidic line), sacred objects (ark and cherubim), sacred sites (Jerusalem and Mount Zion), shrines (tent and temple), priesthoods (Zadokites and Levites), and a sacred calendar (festivals). Many of these institutions were undoubtedly part of the social structure and religious faith of previous periods in Israel's history. Several have their roots in the exodus and conquest–settlement periods; the establishment of the monarchy, however, occasioned radical changes in Hebrew religion.

Social Structure during the Settlement Period

The biblical books of Joshua and Judges present apparently conflicting accounts of the conquest and settlement periods. According to the Book of Joshua, the people, under the leadership of Joshua, invaded the land from the Transjordan region. Through a divide-and-conquer campaign, they rapidly defeated all enemies. The newly acquired lands were then distributed among the various tribes for settlement. But according to the Book of Judges, individual tribes, under charismatic leaders called judges, "freed" specific territories for occupation.

One way to reconcile these two accounts is to make them sequential; the Book of Joshua presents the initial invasion and conquest of major cities in the land, while the Book of Judges tells the story of the settlement of the fertile plains. The entire period, dated ca. 1250–1050 B.C.E., was dominated by brutal warfare. Occasionally, tribes cooperated in battles against formidable foes; Deborah united six tribes against the Canaanites (Judg. 5), and Gideon united four tribes against the Midianites (Judg. 6–9). But usually a tribe engaged individually either in warfare against a foreign enemy or in intertribal conflicts. This situation continued until Philistine aggression forced the Hebrew people to assure their own survival by establishing a monarchy.[4]

This rather idealized portrait of the conquest and settlement of Canaan is probably the fanciful product of the deuteronomistic historians. Recent archaeological discoveries, comparative anthropological studies, and detailed literary analyses of the Book of Judges have suggested a radically new view of the settlement period.

Archaeologists agree that at the beginning of the thirteenth century a cultural change occurred in Canaan. Many fortified cities of the Late Bronze Age were destroyed. Emerging to replace them were village settlements of the Iron Age, some located on the sites of the destroyed cities, but others established in the less densely populated hill country. Agricultural productivity increased in the central highland through deforestation, terracing techniques, and cistern construction.[5] The shift in population centers, however, was accompanied by a decline in the techniques of architectural construction and pottery making. Scholars can only conjecture the exact cause and date when the fortified cities were destroyed. One hypothesis is that they were destroyed when Joshua invaded Canaan; the conquest is dated ca. 1250.[6] But from an archaeologist's perspective, there is no reason to suppose the same forces caused all the destructions. In recent years, therefore, a new theory of conquest and settlement has emerged: the land was seized internally by a peasant revolt of the native population in which small-farm holders and herdsmen overthrew their Canaanite overlords in a war of self-liberation. This theory presupposes that Israel originated not from invading nomads, but from Canaanite peasant farmers who entered into a treaty–covenant relationship with Yahweh.[7]

The exact origins of the Hebrew people who emerged during the settlement period may never be known. In *A History of Ancient Israel and Judah*, Miller and Hayes contend that the evidence supports a complex, varied origin.[8] Some of the Hebrew people may be related to an indigenous population who had lived in Canaan for two millennia. Others may have been pastoralists who moved into the cultivated regions when vegetation became limited in their pastures during the dry summer months. An initially seasonal change of pastures then became a permanent settlement in the thinly populated hill country of Canaan.[9] Another group may be related to the people the Egyptians called *'Apiru*, a class who stood outside the regular social system and possessed no legal status. The Amarna Letters indicate *'Apiru* were a disrupting force in Canaan during the late fifteenth and early fourteenth centuries. Other Hebrew people originally may have been part of the "Sea Peoples" who entered the region from the first half of the fourteenth century down to ca. 1200. The Danuna, a subgroup among the Sea Peoples, may possibly be the tribe of Dan.[10] Still others may have escaped slavery in Egypt and invaded the area from the south. Miller and Hayes believe this blending of various peoples ultimately formed the nation of Israel.

The social structure during the settlement period consisted of the extended family, the clan or village, and the tribe.[11] The family unit (*bêṯʾāb*) extended over several generations and included the patriarch with his wife or wives, his sons with their wives, their sons and wives, and all unmarried children and grandchildren, along with close relatives and slaves. They dwelled together, supporting themselves through agricultural and pastoral activities. The next level was probably the clan (*mišpāḥâ*), consisting of several nearby family units. Marriage occurred customarily between families of the same clan. Some scholars, however, believe the village rather than the clan formed the next social unit after the extended family. The village depended on geographic and cultural proximity rather than kinship relations. The third level of social organization was the tribe. Although the tribe often consisted of clans that were related to one another, it was usually defined by a specific geographic area. But people who were not members of the tribe did live within a tribal area.

The Book of Judges implies that the basic organizational unit was the tribe headed by judges, but the title "judge" appears primarily in editorial passages. But the most important social unit politically was the clan or village that was ruled by a group of elders chosen because of their prominence in the area. These elders were in charge of the local government and the administration of justice (Judg. 8:16; Josh. 20:4; Ruth 4:2). Other officials in the community—called "rulers," "leaders," or "heads"—also may have administered justice, but were primarily military men.

A unified twelve-tribal system is an idealized view of the settlement period from a later age.[12] Miller and Hayes conclude that the tribe of Judah was relatively insignificant. It was located in the south along with such other tribes as the Calebites, Korathites, Simeonites, Kenizites, Jerahmeelites, and Kenites. Ephraim formed a loose alliance with Benjamin, Gilead, and Manasseh, and dominated the north-central hill country. Occupying the far north were the Galilee-Jezreel tribes, a separate alliance of Asher, Naphtali, Zebulun, and Issachar. Miller and Hayes conjecture that the tribe of Dan should be equated with the Danuna, a subgroup of the Sea Peoples who initially occupied territory in the Philistine plane by the Mediterranean Sea. Philistine pressure forced them to relocate at the foot of Mount Hermon. The tribes of neither Reuben nor Gad secured any identifiable land; their people were pastoralists whose grazing ranges overlapped with other tribal territories.[13]

Religion of the Settlement Period

The individual stories of the Book of Judges, along with other biblical and nonbiblical evidence, reveal that the religion of the settlement period was not radically distinct from the Canaanite faith in the area. Worship was conducted at a number of ancient Canaanite cult centers like Bethel, Mizpah, Gilgal, Hebron, Kadesh, Gibeon, Beer-sheba, Nob, Laish (renamed Dan), Penuel, Shechem, and Shiloh.[14] The Book of Judges is strangely silent concerning Shiloh; however, I and II Samuel and Psalm 78 assign this sanctuary a prominent place during the settlement period. The priests who performed rites at these centers apparently came from various tribes, clans, and families. The later prominent, priestly lines of Levites and Aaronites were probably associated respectively with Bethlehem (Judg. 17:7–8; 19:1) and Bethel (compare the golden calves of Exod. 32 with those of Bethel). During the Davidic period Zadokites, associated with the Jerusalem cult, became eminent. Later editors of the biblical material attempted, rather unsuccessfully, to relate all priests to the Levitical line.

The records differ concerning which god was worshiped at these cultic centers. The deuteronomistic editors of the Book of Judges leave the impression that the faithful worshiped Yahweh, the God of Israel. Patriarchal stories and the individual narratives of the Book of Judges, however, indicate that the Canaanite gods El, Baal, and Asherah[15] were honored along with Yahweh. The patriarchs were primarily worshipers of El, the high god of the Canaanite pantheon. Although that god was still worshiped during the settlement period,[16] the cult of Baal and Asherah was more widespread. Apparently Yahwism and Baalism coexisted; both gods were worshiped with similar cultic rites.

While the Hebrew people held much of Canaanite religion in common with their neighbors, they alone worshiped Yahweh.[17] Frank Cross, in his study of the poetry of the Ancient Near East, identifies early Hebrew poems that he believes formed a part of the cultus at Gilgal and Shiloh during the settlement period. In these poems Yahweh is a divine warrior who gives victories in battle to his followers.[18]

We have already considered that in Babylonian culture Marduk led the assembly of the gods in a battle against Tiamat and Kingu, the cosmic forces of chaos. Similarly, Canaanite culture depicted Baal as a divine warrior who overcame the forces of disorder. Although Baal was primarily a dying-and-rising god related to the seasonal cycles, he was also a militant deity who subdued Prince Sea (*Yamm*) and Judge River (*Nahar*) with the help of a double mace named Driver and Expeller. A fragmentary text hints at the dismemberment of *Yamm/Nahar*. The Scandinavian scholar Arvid Kapelrud suspects that fragment probably preceded a fully developed creation myth.[19]

Although stories of El generally picture him as more passive than Baal, El was also an active, divine warrior. The many references to him as "Bull El" have usually been interpreted as allusions to his sexual prowess. But relating the bull to an image of strength and might is just as defensible. The name El is often associated with such concepts as "warrior," "attacking," "striking down," and "a lion," all conveying aggression in battle. In the Keret texts, El aids King Keret in preparing a military expedition to seek a wife.[20] Even greater warlike qualities are attributed to El (under the name Kronos) in Sanchuniathon, a source for "The Phoenician History" of Philo of Byblos:

> When Kronos reached manhood, he punished his father Ouranos and thus avenged his mother, utilizing Hermes Trismegistos—for he was his secretary—as counselor and helper. The children of Kronos were Persephone and Athena. The former died in early maidenhood; with the advice of the latter, Athena, as well as of Hermes, Kronos made a sickle and spear of iron. Then Hermes used magic spells on the allies of Kronos and instilled in them a desire to fight against Ouranos on behalf of Ge. Thus, Kronos waged war against Ouranos, expelled him from his dominion, and took up his kingdom.[21]

The Hebrew people drew upon this Canaanite tradition in developing their imagery of Yahweh as a divine warrior. Whereas Baal and El usually battled cosmic chaos, Yahweh fought for Israel against historical opponents.

Cross detects two divine-warrior traditions that influenced the Hebrew faith of the settlement period. In one tradition, Yahweh comes from Sinai to do battle on behalf of his people.

> When Thou, Yahweh, went forth from Seir,
> When Thou didst march forth from the highlands
> of Edom.
> Earth shook, mountains shuddered;
> Before Yahweh, Lord of Sinai,
> Before Yahweh, God of Israel.[22]
>
> (Judg. 5:4–5)

Similarly, Deuteronomy 33:2–3:

> Yahweh from Sinai came,
> He beamed forth from Seir upon us,
> He shone from Mount Paran.
> With him were myriads of holy ones
> At his right hand marched the divine ones
> Yea, the purified of the peoples.[23]
>
> (See also Ps. 68:7–8;
> Deut. 33:26–29; Hab. 3:3–6)

In the other tradition, Yahweh is the divine warrior who defeated the Egyptians at the Reed Sea[24] and gave Israel the land as a heritage. From the "Song of the Sea" (Exod. 15:1–18), we read:

> Sing to Yahweh,
> For he is highly exalted,
> Horse and chariotry
> He cast into the sea.
> Yahweh is a warrior,
>
> Yahweh is his name.
> Pharaoh and his army
> He hurled into the sea.
> His elite troops
> Drowned in the Reed Sea.
>
> You brought them, you planted them
> In the mount of your heritage,
> The dais of your throne
> Which you made, Yahweh,
> The sanctuary, Yahweh,
> Which your hands created.
>
> Let Yahweh reign
> Forever and ever.[25]
>
> (Exod. 15:1b, 3–4, 17–18;
> see also Hab. 3:7–15;
> Pss. 77:16–20; 114)

What united the Hebrew people in their struggle to possess the land was faith in Yahweh, the divine warrior who overcame all foes. Because Yahweh declared the wars, he must be consulted through numerous cultic methods before battles were fought. God might reveal himself through dreams (Judg. 7:9–14), through casting Urim and Thummim (I Sam. 28:6),[26] through ephod (I Sam. 23:9; 30:7),[27] or through words from a prophet (II Sam. 5:19–24). Once divine approval came, the people prepared for battle by offering to Yahweh a sacrifice (I Sam. 7:9; 13:9, 12).

The commander of the army was readied for his office by the gift of God's spirit (Judg. 6:34). All who served in the army should be totally devoted to God; those who were newly married or had other concerns were excused (Deut. 20:5–9). As soldiers in Yahweh's army, they were to take vows of chastity (I Sam. 21:4; II Sam. 11:11) and rid the camp of all human excrement (Deut. 23:12–14). When the ark was taken into battle, the men would shout, "Arise, O LORD, and let thy enemies be scattered. . . ." When the ark returned from battle, they would say, "Return, O LORD, to the ten thousand thousands of Israel." (Num. 10:35–36).

The sound of the ram's horn assembled the army for the attack (Judg. 3:27); the soldiers shouted the battle cry, "For the LORD has given your enemies . . . into your hand"; the men believed God used the forces of nature to bring victory (Judg. 5:4–5, 20–21). Always the victory was God's. The armies of Israel simply completed what Yahweh had already accomplished (Judg. 4:15–16). Some cities within Israel had the "ban" (*ḥerem*) placed on them and were devoted to God through their own destruction (Josh. 6:17); in this manner, the land was freed from the enemy so that "peace" (*šālōm*) could be established. The holy war was a war of liberation, a war for survival, a war to assure the existence of a people.

Our historical sources concerning these holy wars are largely deuteronomistic and priestly. For preserving the ancient traditions these sources may vary in degrees of accuracy on specific details. But the antiquity of the poems depicting Yahweh as the divine warrior and the corroborating ideology among the Philistines (I Sam. 4:7) and the Moabites (II Kings 3:21–27 and the Mesha Stone) lead contemporary students to accept the accuracy of the sources for expressing the faith of the settlement period.

The Establishment of the Monarchy

The adoption of the monarchy made a formative impression on Hebrew history and religion. I will not trace the steps by which first Saul, then David and Solomon acquired the throne. Rather, I will characterize the social, political, and religious nature of the monarchy as it first emerged in the united kingdom and then as it developed in both Judah and Israel.[28]

Saul, a native of Benjaminite territory, achieved some notoriety when his small private army expelled the Philistines from the central hill country of Canaan. But the incident that led to his "national" prominence was his deliverance of the people of Jabesh in Gilead from the ruthless oppression of their Ammonite neighbors. These two military victories catapulted him into the role of protector and ruler over the people in his region. Traditions vary concerning his

selection as king. According to what scholars call the Early Source
(I Sam. 9:1–10:16; 11:1–15), Saul was anointed[29] by Samuel and
crowned at Gilgal. The Later Source (I Sam. 8:1–22; 10:17–27) calls
Samuel the last judge over Israel who opposed the establishment of
the monarchy. Only when Saul was elected king by lot did Samuel
reluctantly concur. Saul was then crowned at Mizpah.[30] The two
sources reflect the mixed feelings in Israel toward the monarchy. For
some, the king was God's representative, selected to establish justice,
peace, and prosperity in the nation; for others, he was a despot who
would eventually rob the people of their freedom. The monarchical
form of government, however, was part of the Israelite–Canaanite
social world, and its establishment was inevitable.[31]

Saul ruled the territories occupied by the tribes of Ephraim, Manasseh, and Benjamin, and later established influence over the southern area of Judah. We know little about his administration of the
region. His modest court was financed by the spoils of war and by
support from those under his protection. According to I Samuel 22:7,
he instituted a policy of making royal land grants that created loyal
military leaders. These military leaders may have been responsible
for administering justice on their estates, but more likely a village
court system operated during the settlement period.

Saul is portrayed as a Yahwist throughout the biblical material.
Although he named a son Ish-Baal or Esh-baal,[32] the term "baal,"
which means "lord," was a synonym in early Israel for Yahweh.
Through Samuel, Saul had a special relationship with the sanctuary
at Shiloh that housed Yahweh's sacred ark. As a follower of Yahweh,
he engaged in holy wars. Although he did not live to found a state
religion, he was concerned with honoring Yahweh who placed him
on the throne. The deuteronomistic historians credit him with building
the first altar dedicated to Yahweh (I Sam. 14:31–35).

David the Empire Builder

The biblical record is historically more useful in the material concerned with the establishment of the united kingdom of David. The
text is not history in the modern sense; but there is less emphasis on
the legendary, the cultic, and the popular traditions found in earlier
materials. R. N. Whybray has designated as "political propaganda"
portions that reflect the influence of wisdom circles within the royal
court.[33] David M. Gunn believes the account is a *novella*.[34] These

categorizations have led some scholars to question whether the material contains any accurate information about what took place.[35] Other critics regard such skepticism as extreme and contend the narratives are quite useful to historians.[36] This body of literature must obviously be approached with great caution.

David created the largest empire Israel ever possessed.[37] Saul's surviving son Esh-baal inherited the throne,[38] and for several years a rivalry existed between him and the charismatic, popular David. The "men of Judah" (II Sam. 2:1–4) anointed David king of Judah at the ancient shrine of Hebron.[39] Subsequently, after Esh-baal's murder, the elders of the tribes of Israel also anointed David king of Israel (II Sam. 5:1–5). A covenant document, probably stating mutual rights and obligations, sealed the relationship between the northern people and David. Because David had previously married Saul's daughter Michal, he demanded that she be returned to him.[40] Perhaps the newly anointed king was trying to legitimize his claim as Saul's successor. But because Saul was from the northern tribe of Benjamin, David was more likely hoping to unite the north with the south.

One of David's early accomplishments was the capture of Jerusalem. Apparently taken without a major battle by David's personal army, Jerusalem became a part of his private holdings; thereafter it was called "the city of David." Why David chose it as the site for his future capital is unknown. The city was not the most militarily defensible in the land. Some scholars suppose that because Jerusalem possessed the most sacred site in all Canaan (i.e., Mount Zion), David wished to establish the Yahweh cult there. But the pre-Davidic origin of Zion is difficult to substantiate; in fact, the evidence supports a Davidic origin for the Zion tradition.[41] The most likely theory is that the city-state of Jerusalem provided a neutral site favoring neither the north nor the south. Its geographic location on the border between the two rival sections of the country proved useful in David's attempt to unify the new nation. Although Israelites began living in Jerusalem during David's rule, an extensive Canaanite population remained, which influenced extensively David's newly defined Yahweh cult.[42]

David is rightly called the empire builder. The territory ruled by the new nation was extensive. Because Egyptian influence was declining and Assyria had not yet emerged as the dominant power in the region, the time was ideal for the nation of Israel to become a principal power. In addition to the capture of Jerusalem, David apparently gained control over those Canaanite city-states that had previously maintained their freedom from Hebrew domination. How fully they were integrated into the new nation varied. In some instances the Canaanite city rulers would be subject to the crown; in other instances, especially Jerusalem, they would be absorbed by the new nation. This introduced into Israel a class system with wealthy aristocrats foreign to the egalitarian social structure of the settlement period. Israel was going "the way of all nations."

David formed his empire through marriage agreements, conquests of neighboring lands, and treaties with foreign powers. While David was still at Hebron, the small Aramaean state of Geshur in the northern Transjordan region probably became allied with Israel through his marriage to the king's daughter (II Sam. 3:3). David subdued the Philistines by conquest. He had already defeated them near Jerusalem on two separate occasions (II Sam. 5:17–25); apparently he continued to drive them from the land. He finally forced the Philistines into a small pocket along the coast that began south of Joppa and included the cities of Ashdod, Ashkelon, and Gaza.

Ammonite provocation occasioned even further conquests. When David's ambassadors to Ammon were insulted, his army under Joab invaded the area. The Ammonites hired the Aramaean states to the north to protect them, but Joab defeated both armies. David was crowned king of Ammon, but he probably ruled the nation through a local leader. The Aramaean states were punished for intervening in the war. Damascus, their leading city, became a province in David's empire and probably held administrative control over all Syrians (II Sam. 8:3–8; 10:1–19; 12:26–31). David's conquest of the southern Transjordan areas of Moab (II Sam. 8:2, 13–14) and of Edom (I Kings 11:15–18) was especially brutal; Moab became a vassal state and Edom, a conquered province. David further entered into an overlord–vassal treaty relationship with the city-state of Hamath, north of Syria (II Sam. 8:9–10); he negotiated a parity treaty with Hiram of Tyre (II Sam. 5:11–12).[43]

David's empire encompassed an area ranging from the desert in the east to the sea in the west and from the Gulf of Aqabah in the south to Hamath in the north. The Canaanites were incorporated into the state. Edom, Moab, Ammon, and the Aramaean states of Syria came under a provincial administration, while the wealthy city-state of Tyre formed a treaty with David. With the exception of the Philistines along the coast, all these groups shared a similar ethnic, linguistic, and cultural background. They were all emerging independent nation groups. What united them was David himself.

One cannot overemphasize the difference between David's empire and the loosely knit tribal organization of the settlement period. The foundation of this empire was built upon what the German biblical scholar Albrecht Alt called the "personal union" David established between himself and the kingdoms of Judah and Israel. By popular assembly he was made king first of Judah (II Sam. 2:4) and then of Israel (II Sam. 5:1–5). His personal leadership over Israel and Judah was assured through the existence of a private army, under Joab, especially loyal to the king.

The Administration of the State

We have considered the simple governmental structure Saul used to administer his state. David and Solomon developed more elaborate systems. N. K. Gottwald maintains that David organized the twelve-tribe system for political and administrative reasons, while Solomon found it economically and militarily advisable to redistrict the nation.[44] Jerusalem was the central administrative site; the empire became an extended city-state with all officials directly responsible to the king.

II Samuel 8:15–18 and 20:23–26 list the major officials in the Davidic kingdom.[45] A commander over the army was responsible for conducting the wars of expansion. The king also had a personal army consisting of Cherethites (Cretans?), Pelethites (Philistines?), and Gittites (II Sam. 15:18), each group with its own commanders. A recorder kept the official state documents relating to domestic affairs; a secretary probably handled the diplomatic correspondence with other nations. An official administered the "forced labor" used both for public and royal projects.

Miller and Hayes argue that during David's reign the Levites played a prominent role in the administration of the state.[46] Tradition associates them with Bethlehem (Judg. 17:7–8; 19) and Hebron (I Chron. 26:30–32). Miller and Hayes further suggest that the list of Levitical administrative cities (Josh. 21:1–42) comes from the Davidic period. These Levitical cities were established by land grants from the crown.[47] Their locations indicate the approximate extent of David's empire.[48]

According to I Kings 4:7–19, Solomon added to David's bureaucracy two officials, one over the officers and another over the palace. Twelve administrative officers were placed over newly formed districts of the nation; these officers served largely as tax collectors supporting the Jerusalem court. The passage closes with the observation that "there was one officer in the land of Judah." Some scholars regard this verse as editorial and believe Solomon taxed all tribes except Judah. Others hold that the taxes were evenly spread throughout the realm.

According to G. W. Ahlström, royal administration and national religion were intimately related. The founding of cities and fortresses during the Davidic–Solomonic period, most of which had sanctuaries, was for the purpose of administering the territory. "By sending out and placing military personnel and civil servants including priests in district capitals, at strategic points, in store cities, and in the national sanctuaries, the central government saw to it that both civil and cultic laws were upheld and that taxes were paid."[49]

The Ark of the Covenant

Biblical sources describe the religious changes that occurred in the new state. Apparently David established a state religion that attempted to reconcile Canaanite and Hebrew faith; thus he united a diverse people.[50] Honoring his own ancestral traditions, he transferred the sacred ark of the covenant from Kiriath-jearim to Jerusalem.

Scholars debate the antiquity and the meaning of the ark.[51] All pentateuchal traditions trace the ark back to the Mosaic period. The earliest narratives (JE) describe a simple wooden box that represents the presence of God (Num. 10:35–36). In later narratives (D and P), the box contains the two law tablets from Sinai (Deut. 10:1–5; Exod. 25:10–22).[52] During the period of the settlement, the ark was carried into battle to represent God's presence with his people (I Sam. 4–7). It may have been a pedestal on which the invisible Yahweh reputedly stood. If the traditions are accurate, the ark first resided at Bethel (Judg. 20:27–28). But by the close of the settlement period it was at Shiloh; from there it was carried into battle against the Philistines and captured (Judg. 4–7). It may have remained under Philistine control until the time of David.[53] Some scholars believe the ark also resided at Kadesh-barnea before its transport into Canaan by the Hebrews, at Gilgal as a part of an annual celebration of the crossing of the Jordan (Josh. 3–4), and at Shechem.

The Cherubim

Understanding the cherubim enriches one's appreciation of the significance of the ark. Recall that in Canaanite art both the god and his representative, the king, sat upon a throne supported by sphinxes of judgment.[54] These heavenly creatures of the storm clouds represent the divine presence; they symbolize the cloud chariot of the god who rides across the skies. The Yahweh cult adopted this imagery to convey God's presence. Thus Solomon's temple was decorated with cherubim, and the Holy of Holies possessed two cherubim with wings outstretched above the ark (I Kings 6 and 8).

Some scholars maintain that the association of ark, cherubim, and the divine name of Yahweh of hosts first took place in the Jerusalem cult and was read back into the settlement period. But such skepticism concerning the traditions is unwarranted. I Samuel 4:4 reads:

> So the people sent to Shiloh, and brought from there the ark of the covenant of the LORD of hosts, who is enthroned on the cherubim.

This Shiloh passage not only links the ark with the cherubim, but also uses the Jerusalem cult's favorite divine designation, Yahweh of hosts. II Samuel 6:2 also relates that David brought to Jerusalem the ark of God, "which is called by the name of the LORD of hosts who sits enthroned on the cherubim." Shiloh was possibly the origin for the association of ark, cherubim, and the divine name Yahweh of hosts. If so, Shiloh would have been where Yahweh, God of the Exodus, was first proclaimed the divine king Yahweh of hosts, enthroned on the cherubim.[55] What better way for David to relate his earthly kingship both to the Hebrew tradition (the ark) and to the Canaanite culture (enthroned on the cherubim)?

A comparison of II Samuel 6 and Psalm 132:1–10 reveals that the former contains the historical tradition of David's removal of the ark to Jerusalem while the latter presents the cultic celebration of the event.[56] The psalm probably accompanied a dramatic reenactment of the ark's discovery at Kiriath-jearim, where it may have resided since Samuel's day, and of the procession by which the ark was brought into Jerusalem. The psalm recounts David's enthronement in Jerusalem at Mount Zion, perhaps David's last coronation. David was probably made king on three separate occasions: first over Judah, then over Israel, and finally over Jerusalem. His third coronation would have involved the ark, symbolic of the kingship of Yahweh of hosts who sits enthroned on the cherubim. J. R. Porter argues that by bringing the ark to Jerusalem, David instituted a Canaanite-style New Year Festival that became an annual celebration of both the enthronement of Yahweh and the establishment of the Davidic line. This goes beyond the evidence. David was simply providing religious sanction for his kingship.

El Ṣëbā'ôt

The divine name *Ṣëbā'ôt,* "hosts," is associated with the sanctuaries at Shiloh and Jerusalem. The meaning of *ṣëbā'ôt* is ambiguous. The most widely accepted view regards it as a form of the Hebrew word *ṣābā',* meaning "soldier," or "army," hence "Yahweh, God of Israel's armies." Others believe it originally referred to the stars, hence the "hosts of heaven."[57] Still others maintain it refers to the "sons of God" who constituted the Council of Yahweh.[58] And, not uncommonly, scholars embrace all three meanings under the phrase "God of hosts."

A British scholar, J. P. Ross, believes that the term is Canaanite in origin. He maintains that Shiloh once honored cherub-enthroned Baal *Ṣëbā'ôt.*[59] When the Hebrew people occupied that city during the settlement period, they not only housed the ark of the covenant in Shiloh, but they also adopted both the cherub-throne and the term "Lord of hosts" as expressions of Yahweh's majesty. Ross does not think the term originally designated Yahweh as a leader of Israel's armies, but rather conveyed the royal nature of Israel's God. When David established Jerusalem as his capital, he brought to that city the ark of the covenant "called by the name of the Lord of hosts who sits enthroned on the cherubim" (II Sam. 6:2) in order to legitimate his dynasty. Finally, Solomon constructed the temple to house the ark and fashioned two cherubim for the Holy of Holies (I Kings 6). Henceforth, the monarchy in Judah was rooted in the kingship of Yahweh who sits enthroned on the cherubim on Mount Zion.

The Tent of Meeting

David placed the ark in the tent of meeting, joining the two most sacred cultic objects from the wilderness period.[60] These two objects were apparently not associated before David's time. Although the earliest sources place both the ark (Num. 10:33–36; 14:44) and the tent of meeting (Exod. 33:7–11) in the wilderness period, they are never united within the text.[61] According to Exodus 33:7–11, during the wilderness wanderings the tent of meeting was placed some distance from the camp. When Moses would enter the tent, the pillar of cloud, representing God's presence, would descend, coming to rest in the doorway. Out of the cloud, God would speak to Moses. If the ark was in the tent, it is strange that the cloud would appear in the doorway rather than over the ark.

However, I Samuel refers to both the ark (1:9) and a tent of meeting (2:22) at Shiloh. This, then, may be the place where the two were first together; but the relationship among ark, tent, and the Shiloh sanctuary is equivocal. There is no question that the ark was at Shiloh; as an ancient Canaanite cultic center, Shiloh certainly possessed a temple. That the original wilderness tent survived throughout the settlement period is doubtful. David probably had a replica of the tent made to house the ark, and the deuteronomistic historians assumed the tent had been at Shiloh with the ark. David would therefore be the first to bring together these two cultic objects of early Yahweh worship,[62] providing a strong tie binding the new national religion to the faith of the period of the judges.

Shiloh and Jerusalem

At one time scholars claimed that the central sanctuary at Shiloh was the site where the twelve-tribal amphictyonic league gathered annually to renew its covenant with Yahweh. This theory is no longer widely accepted. Nevertheless, Shiloh was the most prominent worship center during the settlement period. Although the law of the altar in the Book of the Covenant indicates there were multiple places of worship in the land (Exod. 20:24), Shiloh eventually overshadowed them all and became the place to which annual pilgrimages were made (Judg. 21:19–21). At Shiloh, Mosaic and Canaanite faith were uniquely united during the settlement period. The tent of meeting probably did not reside at Shiloh; but at an ancient Canaanite sanctuary there, people worshiped the Canaanite God named "El Ṣĕbāʾ ōt who sits enthroned upon the cherubim." When Israel appropriated the sanctuary, the ark was probably placed in the temple and Yahweh was identified with El Ṣĕbāʾ ōt.[63] Shiloh became the first temple in the land dedicated to Yahweh. With the defeat of the Philistines, David formed a united kingdom and established a state Yahweh cult in Jerusalem. By bringing the ark to Jerusalem, he made that city the successor of Shiloh.

Mount Zion

David supported his empire theologically by declaring Mount Zion the unique dwelling place of Yahweh, thus invoking traditions drawn from El and Baal mythology. When David placed the ark and the tent on Mount Zion within Jerusalem, in Hebrew faith Mount Zion became the cosmic mountain of God, the dwelling place of Yahweh.[64] (As we have seen, in the Ancient Near East gods were identified with sacred mountains: El dwelled on a cosmic mountain and Baal on Mount Saphon.) Not all scholars agree that David was the first to declare Mount Zion Yahweh's dwelling place. Although many trace its origin as a cosmic mountain to the pre-Israelite period,[65] J. J. M. Roberts argues persuasively for its Davidic origin.[66]

Roberts designates five Zion motifs, the first three connected with El and Baal mythology, and the last two related uniquely to Israel's history. First, in Psalm 48:1–2, Mount Zion is identified with Mount Saphon:

> Great is Yahweh, and greatly to be praised.
> In the city of our God is his holy mountain,
> The most beautiful peak, the joy of all the earth.
> Mt. Zion in the heights of Zaphon, the city of the
>   great king.[67]

The meaning of the passage is obscured by translating Saphon "north." Saphon came to mean north in the Hebrew language because the sacred mountain of Baal was north of Canaan. This comparison of Mount Zion, Yahweh's dwelling place, with Mount Saphon, Baal's holy mountain, does not mean that Yahweh and Baal were identified. Rather, Yahweh faith was being expressed through Baal imagery.

Second, the rivers of paradise flow out of Zion, bringing fertility and healing to the people.

> There is a river whose stream makes glad
>   the city of God,
> the holy habitation of the Most High.
>   (Ps. 46:4; see also Ezek. 47:1–12;
>   Zech. 13:1; 18:8)

This paradisical river is not in Baal mythology, but El purportedly dwells on the cosmic mountain where two emerging rivers are the sources of the two seas.[68] This is probably the origin of Zion's river of paradise.

Third, because Yahweh dwells on Zion, the primordial waters of chaos are no threat to his people. In praise of Zion, the psalmist writes:

> God is our refuge and strength,
>   a very present help in trouble.
> Therefore we will not fear though
>   the earth should change,
>   though the mountains shake in the
>   heart of the sea;
> though its waters roar and foam,
>   though the mountains tremble with
>   its tumult.
> God is in the midst of her, she shall
>   not be moved;
>   God will help her right early.
>   (Ps. 46:1–3, 5)

As Marduk defeats Tiamat in Babylonian myth, so in Canaanite mythology Baal battles Prince Sea (*Yamm*) and Judge River (*Nahar*), defeating the forces of chaos and bringing order from disorder. Hebrew faith borrowed directly from Canaanite culture to express Yahweh's victory over the chaotic waters.

> The earth is the LORD's and the fulness thereof,
>   and the world and those who dwell therein;

for he has founded it upon the *seas*,
and established it upon the *rivers*.
(Ps. 24:1–2; italics mine)

The final two Zion motifs that Roberts mentions are not connected with Canaanite mythology, but are related to events in Israel's own history. Zion is the center to which foreign nations pilgrimage to acknowledge Yahweh's sovereignty. While this theme is especially prominent in postexilic apocalypticism, Roberts argues that it may be Davidic in origin. For when David established his empire, the surrounding vassal states would be expected to pay him tribute and to honor Yahweh, Israel's national God. These acts would be in keeping with the accepted practices of the nations of the Ancient Near East toward their liege states.[69] The second "historical" element of the Zion tradition is the defeat of the nation's enemies that have gathered to attack Jerusalem.[70] This theme, prominent in the apocalyptic thought of a later day, undoubtedly had some historical basis. Some scholars suggest a parallel in Sennacherib's unsuccessful campaign to take Jerusalem during Hezekiah's reign. Although the Zion tradition is an important part of Isaiah of Jerusalem's prophecy during that period, Roberts proposes that David's wars against the Philistines (II Sam. 5:17–25) shortly after he captured Jerusalem may provide the historical origin for this element in the Zion tradition.[71]

The Davidic Covenant[72]

Moshe Weinfeld's study of "royal grants" in the Ancient Near East reveals the close relationship between David's transfer of the ark to Zion (II Sam. 6) and the divine promise to establish an everlasting dynasty for David (II Sam. 7).[73] Three essential elements of the grants influenced the form and content of the Davidic covenant: the grants were based on the loyalty of individuals to their masters, the unconditional nature of the grants made them everlasting, and the most prominent gifts of the sovereign were land and dynasty.

David's loyalty to God appears not only in the transfer of the ark to Zion, but also in his desire to build a temple for the ark's permanent dwelling place. The latter provides an opportunity for a clever play on the Hebrew word for "house." Because David dwells in a palace (house), he wishes to build a temple (house) for God; but God reveals to Nathan that he chooses to continue dwelling in the wilderness tent. Instead of David's building a temple (house) for God, God will establish for David an everlasting dynasty (house).

> When your days are fulfilled and you lie down with your fathers, I will raise up your offspring after you, who shall come forth from your body, and I will establish his kingdom. He shall build a house for my name, and I will establish the throne of his kingdom for ever. I will be his father, and he shall be my son. When he commits iniquity, I will chasten him with the rod of men, with the stripes of the sons of men; but I will not take my steadfast love from him, as I took it from Saul, whom I put away from before you. And your house and your kingdom shall be made sure for ever before me; your throne shall be established for ever.[74]   (II Sam. 7:12–16)

Nathan's oracle is remarkably similar to a Hittite king's grant to his vassal:

> After you, your son and grandson will possess it, nobody will take it away from them; if one of your descendants sins, the king will prosecute him, . . . but nobody will take away either *his house or his land* in order to give it to a descendant of somebody else.[75]

Even the father–son imagery in God's covenant with David ("I will be his father, and he shall be my son") has its counterpart in the Hittite royal grants.

> [The great King] grasped me with his hand and said: "When I conquer the land of Mitanni I shall not reject you, *I shall make you my son.* I will stand by [to help you in war] and will make you sit on the throne of your father . . . the word which comes out of his mouth will not turn back."[76]

The themes of sonship (II Sam. 7:14; Pss. 2:7; 89:26), grasping the hand (Ps. 89:21), and the word that never fails (Ps. 132:11) are all part of the Davidic-covenant tradition.[77] The Davidic king becomes the "son of God" when he is enthroned. This does not give him divine status, but it does indicate that God has chosen him to rule in a godlike manner.[78] Henceforth, God is a Father to the king, and the king assumes the duties and privileges of a son.

Psalm 132 contains this same connection between the transfer of the ark to Zion and the establishment of a dynasty, except that the

Davidic covenant in Nathan's oracle is everlasting (unconditional), while in Psalm 132 it is conditioned by the faithfulness of the king. Psalm 89 expresses the Davidic covenant in a form identical to Nathan's oracle.[79]

While the Abrahamic and Davidic covenants have long been recognized as typologically similar,[80] the three major elements found in the royal grants and the Davidic covenant are also, clearly, essential parts of God's covenant with Abraham. Because of Abraham's faithfulness, God promised to grant him descendants and land, promises that would never be withdrawn. The Yahwistic expression of this covenant (Gen. 15) was formulated during the united monarchy. Both covenant traditions took shape during the same period. The Davidic covenant emphasized the establishment of a dynasty that cannot exist apart from the land over which David ruled. Conversely, the Abrahamic covenant emphasized land, a promise that has no meaning apart from descendants (dynasty?). Significantly, the land promised to Abraham (Gen. 15:18–21) is identical to the boundaries of the Davidic empire.

David's initial intention was to build a temple honoring the God who had placed him on the throne. This was consistent with the customs of the Ancient Near East. Probably behind Nathan's oracle opposing the building of a temple was a genuine resistance within Israel to the substitution of a Canaanite-style temple for a wilderness-style tent. Although Canaanite shrines were being used as Yahweh cult centers, the projected Jerusalem temple would be on a much grander scale. In building a temple, David would be introducing Israel to a dynastic-style monarchy that was foreign to the settlement period. The command not to build a temple may reflect David's awareness that he had introduced into Israel all the Canaanite practices the people would tolerate.

The Role of the King

In the newly created state, the king assumed a prominent role in the worship of Yahweh. On the day of his coronation, the king was sanctified through an anointing ceremony and was declared the adopted "son of God." As a sacred person, he was empowered to perform certain religious functions. On several occasions the king served as a priest, personally performing cultic acts. Saul offered sacrifices at Gilgal (I Sam. 13:9–10), and David officiated at Jerusalem

(II Sam. 6:13, 17–18). Although some texts imply that the king only had priests offer sacrifices on his behalf, one reference shows King Solomon serving as a priest:

> Then he came to Jerusalem, and stood before the ark of the covenant of the LORD, and offered up burnt offerings and peace offerings, and made a feast for all his servants.   (I Kings 3:15)

Furthermore, both David and Solomon blessed the people in the sanctuary (II Sam. 6:18; I Kings 8:14), a rite normally performed by priests. The king usually functioned as a priest only on special occasions, not as a regular member of the priesthood; but as part of a state religion in the culture of the Ancient Near East, the king appropriately officiated at major cultic events.[81]

Yahweh as a divine warrior operated chiefly through the king, who was commander-in-chief of the army. Holy-war ideology had made the transition from the settlement period to the monarchy. What were once primarily wars for survival became expansionist wars for carving out an empire among the nations. Citizens were now drafted into the army (David's census was probably for military conscription; see II Sam. 24) to fight unpopular wars. Just as Baal and El fought cosmic battles to bring order out of chaos, so Yahweh's holy wars were for the purpose of establishing law and order. God, the divine warrior, battled against wicked enemies of the state in order to assure that the nation might possess a just peace.

Another major function of the king as God's representative was to establish justice in the land.[82] In keeping with Mesopotamian and Canaanite practices, the king held no legislative powers; he could not enact laws, but he did hold judicial power. There are multiple examples of court scenes in the scripture in which the king rendered verdicts. Solomon's palace contained a "porch of judgment" (I Kings 7:7) where cases were tried. Most local cases would have been settled by city elders, leading citizens of local towns; however, there were professional judges in Israel who were personally appointed by the crown.[83] Absalom's revolt against David reveals the central place the administration of justice occupied in Israel's monarchy. Absalom accused David publicly of failing to provide prompt judgments in the court. The hearts of the people were won when Absalom exclaimed: "Oh that I were judge in the land! Then every man with a suit or cause might come to me, and I would give him justice" (II Sam. 15:4).

It is doubtful that the king participated in a "sacred marriage" ceremony where the fertility of the land depended on his potency. Yet there are two scriptural references to David that could be interpreted within a fertility cult context. When David brought the ark to Jerusalem he exposed himself in a festive dance that his wife Michal considered lewd.[84] She chided him satirically:

> How the king of Israel honored himself today, uncovering himself today before the eyes of his servants' maids, as one of the vulgar fellows shamelessly uncovers himself.   (II Sam. 6:20)

David's reply is intriguing:

> I will make myself yet more contemptible than this, and I will be abased in your eyes; but by the maids of whom you have spoken, by them I shall be held in honor.   (II Sam. 6:22)

What this means is clarified in the next verse:

> And Michal the daughter of Saul had no child to the day of her death.

David refused to have sexual relations with Michal, but he proved his potency with the maids of the court. The second reference occurs when David was old and chilled. The beautiful Abishag was brought to him, but the king did not have sexual relations with her (I Kings 1:1–4). The next verse may provide an appropriate comment on the episode:

> Now Adonijah the son of Haggith exalted himself, saying, "I will be king."

Did Adonijah declare himself king because David could no longer perform sexually?

There are certainly royal psalms that relate the king to the fertility of the land.

> May he [the king] be like rain that falls
> on the mown grass,
> like showers that water the earth!

May there be abundance of grain in the land;
on the tops of the mountains may it wave;
may its fruit be like Lebanon;
and may men blossom forth from the cities
like the grass of the field!

(Ps. 72:6, 16)

The origin of such concepts is undoubtedly the Canaanite fertility cult.

The Priesthood and the Temple

No state religion perpetuates itself without cult officials. David, therefore, appointed two priests to supervise worship: Abiathar and Zadok. The choice of Abiathar was obvious. He was the sole survivor of Saul's frightful slaughter of the citizens of Nob who had befriended David. Abiathar joined David in the wilderness, bringing with him the ephod, a sacred oracle that David consulted on several occasions to discover God's will. Abiathar's loyalty to David was unquestioned, for he had remained with him throughout his "outlaw years." When Solomon exiled Abiathar after David's death, the deuteronomistic historians interpreted the exile as a fulfillment of God's word spoken against the house of Eli at Shiloh (cf. I Kings 2:27 with I Sam. 2:27–30 and 3:11–14). They believed Abiathar was a Shilonite priest. If this was true, then David would have been continuing his policy of making his state the preserver of the sacred institutions of the past. For more conservative Israelites, Jerusalem—now containing ark, tent, and Abiathar as priest—was a second Shiloh. Abiathar's Shiloh connection, however, may have been the result of later editing.[85]

The choice of Zadok raises perplexing problems. Many biblical scholars believe that he was a priest of Jebusite Jerusalem, a worshiper of El Elyon.[86] As David introduced Israel to a Canaanite-style kingship, he also chose a Canaanite priest to help conduct worship. Moreover, Canaanite influence on the Hebrew cult was occurring long before Zadok was made priest; even the sacrificial system that developed in Israel was based on Canaanite practices. Designating a Jebusite as priest over the new state religion would be an additional way to integrate the Canaanite population of Jerusalem with the Israelite conquerors.

Contradicting this theory, I Chronicles 6:4–8, 50–53; Ezra 7:2–5; I Esdras 8:1–2 and II Esdras 1:1–3 all maintain that the Zadokite priesthood traced its lineage back to Aaron of the Levitical line. Although most scholars regard these genealogies as inaccurate,[87] Frank Cross argues that Zadok, a priest from Hebron claiming descent from Aaron, was in opposition to Abiathar of Shiloh, who also claimed to be of the Levitical line through descent from Moses.[88] Both Zadok and Abiathar would then be legitimate Levitical priests. Cross' argument is carefully reasoned but has not gained wide acceptance.

Originally the priesthood was directly under the control of the monarchy. Priesthood was not restricted to a particular tribe, clan, or family. Apparently the king could appoint to the priesthood whomever he wished. While Abiathar and Zadok may have been in the Levitical line, David's sons also served as priests (II Sam. 8:18). Solomon later expelled Abiathar and made Zadok chief priest in Jerusalem (I Kings 2:26–27, 35). Throughout Israel's history, the priests usually supported those in power.

Pre-Israelite Jerusalem undoubtedly possessed shrines dedicated to the local Canaanite deities. Both the gods El Elyon and Shalem probably had their respective temples. Although the Yahweh cult of Shiloh took over the shrine of El *Ṣĕbāʾōt*, a similar takeover did not occur at Jerusalem. Even if Yahweh became identified with El Elyon, no evidence exists that a Canaanite shrine was used for his worship.[89] David did desire to build a temple for Yahweh. But his son Solomon was left the task of constructing a Canaanite-style temple (I Kings 5) on the site purchased earlier by David (II Sam. 24:16–25).[90]

The Holy Land

With the establishment of the monarchy and the temple, Israel declared Canaan to be Yahweh's possession. Zion was the holy mountain, Jerusalem the holy city, and Canaan the holy land. Yahweh dwelt in his temple on Mount Zion and ruled his people through his earthly representative, the Davidic king.

The origin of Israel's ownership of the land as a gift from Yahweh may be traced to the establishment of the monarchy; the view that the land is Israel's by right of conquest may be a later idealization of the early history. Psalm 78 recounts Israel's flight from Egypt, wilderness wandering, and conquest of Canaan. The psalm reaches its climax with the establishment of Mount Zion as Yahweh's dwelling place, with Israel's possession of the land as Yahweh's gift, and with the choice of David as its ruler.[91]

Rival State Religion in the Northern Kingdom of Israel

At the death of Solomon, the united kingdom was replaced by two rival states. Solomon's oppressive policies of forced labor and discriminatory taxation had alienated the northern tribes. They expected redress from Solomon's heir, Rehoboam, before they would recognize him as king. When Rehoboam foolishly boasted that he would rule more ruthlessly than his father, the north seceded from the union to form an independent kingdom of Israel under the leadership of Jeroboam. This was really not a division of the united kingdom, but rather a refusal of the northern tribes to renew their union with Solomon's successor. The north was simply reasserting the independence it had temporarily relinquished during the reigns of David and Solomon. This withdrawal of the northern tribes marked the beginning of the dissolution of the Davidic empire. Soon those foreign territories under David's rule would also regain their independence.

Retaining the old tribal name of Judah, the south continued to be ruled by those of the Davidic line. Judah always regarded the division of the kingdom as a rebellion and hoped for an early reunification of the nation. Judah inherited all the sacred institutions of the united kingdom—Mount Zion, the ark, the cherubim, the temple, legitimate priesthoods, and a divinely appointed monarchy.

If the northern kingdom was to survive, Jeroboam and his successors would need to unite the people around a rival, state religion.[92] Tirzah was designated the capital, although Shechem and Penuel may have served that function at first (I Kings 12:25).[93] To counteract the prestige of Jerusalem, Jeroboam established an independent Yahweh cult at the ancient Canaanite shrines of Bethel[94] and Dan. Bethel served as a pilgrim site for the southern half of his kingdom, while Dan provided a center for worship in the north. In place of the ark and cherubim, which served as pedestals for the enthronement of Yahweh in Jerusalem, Jeroboam had two golden calves (bulls) built and placed one in Bethel and one in Dan. While these calves were Canaanite in origin (as were the cherubim in Jerusalem), the worshipers probably regarded them merely as supports upon which the invisible Yahweh stood. Certainly Jeroboam was not a Baal worshiper, although often Yahweh was worshiped in the same manner as Baal.[95] Jeroboam showed his independence from the Jerusalem cult by appointing a new, non-Levitical priesthood (I Kings 12:31) to officiate at the shrines.[96] This priesthood remained under the control of the king, who also maintained an official position in the cult with the right to perform sacrifices (I Kings 12:33). Even the calendar of holy days was changed; Sukkoth, the Festival of Ingathering, was held one month later than in the south (I Kings 12:32).[97]

According to the deuteronomistic historians, Jeroboam led the north astray by partitioning the monarchy, by establishing a rival capital to Jerusalem, and by erecting Baal golden calves at Bethel and Dan. Actually Jeroboam was establishing a state religion comparable to that in the south. In place of a Davidic monarchy supported by the Jerusalem cult, he founded a northern dynasty with its own cultic traditions. Initially the northern monarchy remained as loyal to Yahweh as the Davidic line. This situation later changed radically.

4

Amos' Oracles against Foreign Nations

IN HIS CONFRONTATION with Amaziah, Amos insisted that Yahweh called him to prophesy against Israel (7:10–17). There is no indication in either the autobiographical or the biographical texts in the Book of Amos that this Judahite prophet felt compelled to speak to any other people. Is it not strange, then, that the first two chapters of the Book of Amos contain a series of oracles against neighboring foreign nations[1] as well as an oracle against his own kingdom of Judah? Further complicating matters, the remainder of the book contains not one criticism of foreign nations and only two half-verse attacks on Judah (3:1b; 6:1a). How are we to account for this initial attention given to the oracles against foreign nations? Why did the compilers choose to open the Book of Amos with these oracles rather than with Amos' main message against Israel?

Before addressing this issue, we must examine whether all the oracles against foreign nations should be attributed to Amos, and what form the oracles assume. This requires a familiarity with 1:3–2:5.

Authenticity of the Oracles against Foreign Nations

The two criteria used in determining the genuineness of the oracles found in 1:3–2:5 are stylistic and historical. We must decide if the oracles are in a literary style attributable to Amos or if they disclose the hand of later redactors. Are they historically compatible with an eighth-century B.C.E. setting, or do they better denote a later age?

Almost all recent Amos commentaries agree that the oracle

55

against Judah (2:4–5) is a deuteronomistic addition.[2] The usage of the words "law" and "lie" reveals a deuteronomistic viewpoint. The law that is rejected (2:4) reflects a later deuteronomic codification of Israelite legal traditions rather than suggesting the oral instruction of priests and prophets as in the time of Amos (see Hos. 4:6 and Isa. 1:10). Lies, characterized by the phrase "after which their fathers walked," are the idols that have led them astray.[3] To reject the law and to follow false gods define the essence of sin for the deuteronomistic writers. The passage bears the mark of the deuteronomists in both vocabulary and style.[4]

Seventh-century Judah faced the same choices that had confronted Israel a century earlier. The message of Amos was consequently reformulated for this new situation. These revisions by later editors should not be viewed as distorting the original message, but rather as making relevant God's word for a new age. Only in this way does the text become the "living Word of God," eventually accepted as Sacred Scripture by the worshiping community.[5]

Many critics also regard the oracles against Tyre (1:9–10) and Edom (1:11–12) as later additions to the text.[6] Stylistically, these oracles vary in three ways from the other oracles against foreign nations:

1. The oracles against Damascus (1:3–5), Philistia (1:6–8), Ammon (1:13–15), and Moab (2:1–3) contain details concerning punishment beyond the strereotypical "fire" that "devours." The oracles against Tyre and Edom (and also Judah) do not.
2. The oracles against Damascus, Philistia, Ammon, and Moab close with the refrain "says the LORD" (or "says the LORD God," 1:8). The oracles against Tyre and Edom (and also Judah) have no closing refrain.
3. The oracles accuse Philistia and Tyre of committing the same crime (1:6 and 1:9), selling a whole people into slavery to Edom. This type of uncreative repetition is uncharacteristic of Amos.[7]

The same critics also argue that there are historical reasons for dating the oracles against Tyre and Edom in the postexilic period. The strongest case regards the oracle against Edom. After the fall of Jerusalem to the Babylonians in 587, Edom subjected her weakened neighbor to constant military harassment. This conflict produced a number of anti-Edom texts similar to the Amos oracles.[8] The case against a preexilic date for the Tyre oracle is not as strong; it usually rests on similarity to postexilic texts that refer to that city's engage-ment in slave trade.[9] However, evidence for a postexilic date for the Tyre and Edom oracles is not persuasive. Arguments supporting the authenticity of these oracles are presented later in this chapter.

Form of the Oracles against Foreign Nations

The Book of Amos is the first prophetic writing to include oracles against foreign nations, but this type of oracle did not originate with Amos. Duane L. Christensen places its origin in the war oracles used by Israel while conducting holy wars during the wilderness and set-tlement periods.[10] The prewriting prophets, before Amos, trans-formed these war oracles into judgment oracles, usually directed against Israel's foreign enemies.

The original war oracles appear in the cultic rites associated with the conduct of holy wars. The most famous example is the fourth oracle of Balaam that includes a victory blessing for Israel and a curse of defeat for the foreign countries of Edom and Moab.

> Utterance of Balaam, who is Beor's son,
> Utterance of the man whose eye is true;
> Utterance of one who knows the words of El,
> And knows the knowledge of Elyon;
> He beholds visions of Shaddai,
> In a trance, with eyes unveiled;
>
> I see, but not this moment;
> I gaze, but it is not soon—
> When the star of Jacob shall prevail;
> And the scepter shall arise in Israel,
> He shall smite the extremities of Moab,
> And destroy all Bene-Shut.
>
> Edom shall be dispossessed;
> Dispossessed shall be Seir,
> Jacob shall rule over his foes,
> And Israel shall have success;
> Yea, he shall slay the remnant of Ar.[11]
>
> (Num. 24:15–19)

What was initially a cultic, victory–defeat oracle preceding a battle became, in the hands of the prewriting prophets, a judgment oracle.

For example, during the reign of Ahab of Israel (869–850), Ben-hadad of Syria attacked the land and belittled the armies of Israel. This prompted an unnamed prophet to proclaim, in Ahab's presence, a judgment oracle against the Syrians:

> Thus says the LORD, "Because the Syrians have said, 'The LORD is a god of the hills but he is not a god of the valleys,' therefore I will give all this great multitude into your hand, and you shall know that I am LORD."
> (I Kings 20:28)

Although these prophetic words are given in prose form, the overall basic structure is identical to the poetic form of Amos' oracles against the foreign nations (see Table).

Amos' oracles against foreign nations, like the cultic war oracles, were not proclaimed in the presence of the nations to which they were addressed. Instead, the nations are referred to in the third person, never in the second person. Not until the latter half of the oracle against Israel (2:13; the authenticity of 2:10–12 is disputed) does the second-person address occur. This oracle was undoubtedly proclaimed in Israel, probably at Samaria.

Further evidence that Amos used holy-war imagery in proclaiming the defeat of foreign nations is his repeated use of punishment by consuming fire (1:4, 7, 10, 12, 14; 2:2) and his employment of such additional holy-war terminology as "shouting" (1:14; 2:2), "tempest" (1:14), "whirlwind" (1:14), "sound of trumpet" (2:2), and "roaring" (1:2).

To these transformed war oracles, Amos added, perhaps out of the wisdom tradition, the use of graduated numbers ("For three crimes and for four").[12] The numbers are not to be taken literally. Only the final and worse crime is listed, but that crime provides ample justification for the punishment that follows.

Intent of the Oracles against Foreign Nations

We return to the question raised at the beginning of this chapter. What was Amos' intention in proclaiming these oracles against foreign nations? Scholars have offered a variety of answers. There are five possible theories that attempt to explain the place of these oracles in the message of Amos. Though not mutually exclusive, they can be treated separately.

Comparison of Judgment Oracle with Amos' Oracle

I Kings 20:28	Amos 1:3–5

The message formula

Thus says the LORD,	Thus says the LORD:

The indictment

"Because the Syrians have said, 'The LORD is a god of the hills but he is not a god of the valleys,'	"For three crimes of Damascus and for four, I will not cause it to return; because they threshed Gilead with threshing sledges of iron.

The announcement of punishment[a]

therefore I will give all this great multitude into your hand, and you shall know that I am the LORD."	So I will send a fire upon the house of Hazael, and it shall devour the strongholds of Ben-hadad. and cut off the inhabitants from the Valley of Aven, and him that holds the scepter from Beth-eden; I will break the bar of Damascus, and the people of Aram shall go into exile to Kir."
	(author's translation)[b]

[a]Following Mays' analysis of the form of Amos' oracles against foreign nations in *Amos*, p. 23. All that is lacking is the concluding messenger formula, "says the LORD."
[b]Rearranged to restore parallelism.

1. Amos was a monotheist.
2. Amos was a universalist.
3. Amos employed the standards of an international law.
4. Amos employed the style of Egyptian Execration Texts.
5. Amos supported a reunited Davidic empire.

Amos as a Monotheist

To the casual reader, Amos might appear to be a monotheist. In the opening chapters he held six foreign nations responsible for their conduct, as if Yahweh were the God of those nations. The text in 9:7b claims that Yahweh gave the Philistines and the Syrians their present lands, as he gave Canaan to Israel. Few biblical scholars, however, would interpret these verses as monotheistic. Monotheism is the belief in and worship of one God along with the disbelief that any other gods exist. This position is not found in the Book of Amos. Most biblical scholars hold that monotheism was first clearly stated in the exilic and postexilic periods, especially in the writings of II Isaiah (see Isa. 43:11–13; 45:5–6, 18–19, 21).

Three incidents, recorded by the deuteronomistic historians, reveal that Yahweh was understood as a national God in preexilic times. When King Saul found David a dangerous rival, David fled into Philistia, protested his innocence, and said:

> If it is the LORD who has stirred you up against me, may he accept an offering; but if it is men, may they be cursed before the LORD, for they have driven me out this day that I should have no share in the heritage of the LORD, saying, 'Go, serve other gods.' Now therefore, let not my blood fall to the earth away from the presence of the LORD; for the king of Israel has come out to seek my life, like one who hunts a partridge in the mountains.   (I Sam. 26:19b–20)

The historical accuracy of the account does not concern us. What the passage reveals is that Yahweh was identified with a particular nation and a particular land.[13] To leave Israel was to leave the presence of God, was to lose one's share in the heritage of the Lord. Yahweh could not be worshiped beyond the borders of Israel.

A second incident occurred when Jehoram of Israel, Jehoshaphat of Judah, and the king of Edom formed an alliance against the king of Moab. When the Moabite king saw that the battle was going against him,

> Then he took his eldest son who was to reign in his stead, and offered him for a burnt offering upon the wall. And there came great wrath upon Israel; and they withdrew from him and returned to their own land.   (II Kings 3:27)

The phrase "great wrath" is a common way in the Ancient Near East to refer to a deity's act of punishment. The passage indicates that the allies believed they were defeated in the land of Moab because the Moabite god, Chemosh, was sovereign there. They retreated to the safety of territories controlled by their own gods.

A third incident occurred when Naaman, a Syrian, was healed of leprosy by Elisha. Naaman supposedly made a monotheistic confession of faith when he said, "Behold, I know that there is no God in all the earth but in Israel" (II Kings 5:15b). Yet because Naaman wished to continue worshiping God when he returned to Syria, he requested that

> there be given to your servant two mules' burden of earth; for henceforth your servant will not offer burnt offering or sacrifice to any god but the LORD.   (II Kings 5:17)

Although Naaman declared Yahweh was the only God who exists, Naaman could only worship him on Israelite soil.[14]

Maintaining that Amos intended for all nations to acknowledge Yahweh as their God or to deny the existence of their own national deities goes beyond the evidence.

Amos as a Universalist

A related position regards Amos as a universalist.[15] In keeping with Hebrew thought, the emphasis is not placed on Yahweh's metaphysical nature but on his activity in history. This position argues not that Amos was a theoretical monotheist, but rather that he was a prophet who proclaimed Yahweh's universal activity within the histories of all nations.

N. K. Gottwald, a prominent exponent of this position, uses as supporting texts Amos' oracles against foreign nations.[16] He maintains that Yahweh's justification for holding these six foreign nations responsible for their crimes is legitimate only if Amos believed that these nations should have recognized God's sovereign control over them.

Yahweh condemns three of the six nations, however, for what they have done to his own people of Israel. The crime of Damascus and Ammon is their ruthless treatment of God's people residing in Gilead, the Transjordan region of Israel. Israel is also probably the "brother" Edom continuously pursued with a sword. These three oracles, then, are not necessarily universalistic; they could support the view that Yahweh, because he is the national God of Israel, avenges his people against their enemies.

Against whom Philistia and Tyre committed their crimes is more difficult to determine. They are both accused of slave trading, but the identity of the "whole people" sold to Edom (1:6, 9) is unknown. The most likely people geographically are either Israelites or Judahites, but this identity is uncertain. If those enslaved were members of the Hebrew nation, the texts still may not support a universalist position. As in the case of the oracles against Damascus, Ammon, and Edom, Yahweh could be condemning Philistia and Tyre for enslaving his own people. On the other hand, if the enslaved people were non-Hebrews, these verses are open to a universalistic interpretation. The texts remain ambiguous; Gottwald is not justified in using them as unqualified support for his universalistic position.

What remains is the single oracle against Moab. Gottwald believes that in this instance Amos clearly expressed his universalism. The offense of Moab is the desecration of the corpse of Edom's king. No crime has been committed against Israel. Gottwald argues that because this act offended Yahweh, the God of Israel, Amos must consider Yahweh ruler over other nations besides Israel. This is Gottwald's strongest argument supporting a universalistic interpretation of the oracles against foreign nations. However, as we shall see later in this chapter, even this oracle is open to a nationalistic interpretation. The evidence is not as conclusive as Gottwald maintains.

Gottwald further cites three related passages in Amos that he argues support a belief in Yahweh's universal rule over all people.[17] In 3:9–11, Amos depicted Yahweh assembling Assyria[18] and Egypt to witness the punishment of Israel.

> Proclaim to the strongholds in Assyria,
>   and to the strongholds in the land of Egypt,
> and say, "Assemble yourselves upon the
>   mountains of Samaria,
>   and see the great tumults within her,
>   and the oppressions in her midst."
>
> (3:9)

Gottwald contends that the nations must have possessed some agreement on what is ethically correct in order to be aware of Israel's iniquity. Such reasoning is hardly convincing, because it can be argued with equal validity that this passage is merely a poetic device to indicate that Yahweh will make a public spectacle of Israel.

In the second passage, 6:1–3, those who are at ease in Zion and Samaria are asked to compare their nations with Calneh, Hamath, and Gath. Gottwald dates these verses within the ministry of Amos and suggests they refer to the prosperity of these three states during a period of Assyrian relapse. By comparison, these foreign nations are superior to Israel and Judah. When Gottwald cites these verses to support universalism, he implies that Yahweh has made them prominent. Most critics, however, maintain that these verses refer to the conquests of either Tiglath-pileser III (738) or Sargon II (720 and 711) or both.[19] Furthermore, there is no indication that it was Yahweh who brought these three foreign city-states to ruin.

The third passage in support of universalism is 9:7:

> "Are you not like the Ethiopians to me,
> O people of Israel?" says the LORD.
> "Did I not bring up Israel from the land
> of Egypt,
> and the Philistines from Caphtor
> and the Syrians from Kir?"

Gottwald comments:

> There can be no doubt of the importance of this verse. Yahweh not only regards all the nations as similar in his sight ("Are you not as the Nubians to me, O people of Israel?") but extends his concern to the whole history of people so as to include the migrations and present holdings of the Philistines and the Syrians.[20]

Many critics agree with Gottwald[21] and accept this verse as his strongest supporting argument for Amos' belief in Yahweh's supreme power over all nations. A radically different interpretation of this verse will be developed later in this chapter.

The universalism presented in A. S. Kapelrud's book, *Central Ideas in Amos*, resembles monotheism much more than does Gottwald's position. In commenting on the oracle against Moab and 9:7, Kapelrud writes:

> This is a clear break with the idea of national gods for the different peoples. It means that Amos rejects Kemosh as the national deity of Moab. The god "who brought the Moabites" was Yahweh. They were under his command, they had to obey his words.[22]

Kapelrud asserts that the origin of this monotheism is the Canaanite god El, or more specifically his manifestation at Jerusalem as El Elyon. Citing texts that present El as the head of the Canaanite pantheon, Kapelrud follows the German scholar Otto Eissfeldt when he interprets these to mean not only that El is the supreme god, but alse "*the* God" (Kapelrud's italics, p. 44). When David made Jerusalem the religious center of the united kingdom, Yahweh was identified with El Elyon; the result was Hebrew monotheism. In more recent years, however, the work of Frank Cross indicates that Canaanite thought was more polytheistic than monotheistic.[23] Few scholars today follow Kapelrud in rooting Hebrew monotheism in Canaanite culture.[24]

Amos and the Standards of International Law

In 1980, John Barton, an English scholar at Oxford University, published a book on Amos' oracles against foreign nations. In his book he argues that Amos condemned Israel's neighbors for atrocities in war by "appealing to a kind of conventional or customary law about international conduct."[25] Barton does not deny that Israel is held to a stricter accounting under covenant law, a type of revealed law. But he does believe that there existed some accepted norms of international conduct that formed a natural law governing behavior among nations. These standards of conduct were operative in peace and in war. His analysis of international law in the Ancient Near East is arranged under three categories: international law proper (especially treaties), agreed international conventions not legally ratified, and unilaterally accepted norms of international conduct.[26] The entire subject of an ancient international law is both difficult and elusive, far more dependent on inference than on hard evidence. By Barton's own admission, the dividing lines among his categories are uncertain and the texts, when they exist, are often ambiguous. Nevertheless, the theory is intriguing and calls for further study.

Amos and the Egyptian Execration Texts

In recent years, the most widespread approach to these oracles in Amos is to regard them as modeled after the Egyptian Execration Texts. This view was first advanced in a paper read in 1950 by the Danish scholar Aage Bentzen at an international meeting of Old

Testament studies in Leiden and published that same year in *Oud-testamentische Studiën*.[27] The use of the Egyptian Execration Texts to curse the enemies of the state was a common practice in the Middle Kingdom of Egypt. The names of enemy nations and their rulers were inscribed on pottery bowls or written on papyrus and placed in glass bottles that were then smashed. This procedure, the Egyptians believed, diminished the power of the enemy through magic. The nations to be cursed were always listed in a fixed geographic order: first those to the south then those to the north and finally those to the west. A final curse was pronounced against Egyptians; of course, this curse was not against the entire land or its ruler, but against individual criminals. In this way Egyptians purged their land of both external and internal enemies.

Bentzen notes a similar geographic pattern in Amos' oracles against foreign nations: Damascus (northeast), Philistia (southwest), Tyre (northwest), Edom–Moab–Ammon (southeast), then Judah and Israel. He does not suggest any literary dependence of Amos on Egyptian sources nor any conscious borrowing. He simply observes a similar pattern.

The paper suggests that Amos might have delivered these oracles in Bethel at the autumn festival, the northern kingdom's New Year Festival, that celebrated God's triumph over the forces of chaos when he created the world. If this were true, then the enemies of the state would now be regarded as incarnations of the chaotic cosmic forces and would once again be subdued by Yahweh. History and mythology would be united through cultic drama.

Bentzen's position has been enormously popular, even among those scholars who do not accept his myth and ritual approach to Hebrew faith. Both Kapelrud and Gottwald adopt it in their discussion of Amos' oracles against foreign nations.[28] But further study of biblical material and the Egyptian Execration Texts reveals that the similarity between the two has been overstated.[29] Criticism of Bentzen's theory has occurred on both fronts. An examination of oracles against the foreign nations in Amos, Isaiah, Jeremiah, Ezekiel, and Zephaniah exhibits no repeated geographic pattern; the sequence of points on the compass varies in detail. A more thorough study of the Egyptian Execration Texts reveals, furthermore, that the order of south, north, west, and Egypt is a conventional sequence found in *all* Egyptian documents that allude to the geography of the region; the pattern is not unique to the Egyptian Execration Texts. Therefore, the geographic order in the Egyptian Execration Texts has no overt magical design, but is simply the way in which Egyptians ordered their world. The theory that Amos' oracles against foreign nations exhibit a pattern similar to those of the Egyptian Execration Texts is no longer widely accepted.[30]

Amos' Support for a Reunited Empire

In the previous chapter we saw that David made Israel an empire among the nations of the world. Through military conquests, treaties, and marriages, he extended the united kingdom of Israel from Hamath in the north to Elath in the south, from the Mediterranean in the west to the desert in the east. The land bridge between Assyria and Egypt was the Davidic empire.

David personally united this empire, a situation the Hebrew people understood in terms of their theocratic tradition. Yahweh had placed David on the throne; the people stood united under Yahweh. Thus, during this period a state religion emerged that called for the acceptance of God's sovereignty as manifested in the Davidic ruler.[31]

Israel never forgot the glory of David's empire. Even after the destruction of Jerusalem by the Babylonians in 587, hope remained that a king of the Davidic line would regain the throne. In the postexilic period, when the restored peoples were a struggling community, hope for a better day found expression in the belief that God would once again raise up a ruler, a Messiah, from the line of David, who would restore the nation to world prominence.

A recurring problem in the study of preexilic Hebrew prophecy is whether passages dealing with the restoration of the nation are later additions to the text, reflecting the hopes of the postexilic period, or whether they are the genuine beliefs of the prophets to whom they are attributed. The issue is not trivial. Whichever position is taken will affect radically the interpreter's view of the prophet being studied.

In a penetrating article, the British scholar John Mauchline has explored references to the Davidic empire within Hebrew Scripture. Beginning with Amos' oracles against foreign nations,[32] he finds that the only common thread binding together these nations was that all except for Tyre were once ruled by David. In the case of Tyre, a treaty of peace had been formed first with David (II Sam. 5:11) and then with Solomon (I Kings 5:1–18 [H 5:15–32]). These nations, headed by the united kingdom of Israel, were once members of an association of people accepting a mutual covenant. This explains why the crimes of the nations that Amos condemned were international in scope. They were not crimes of theft, or murder, or injustice against citizens of the same country, but were crimes committed among member nations of the Davidic empire. Therefore, Moab could be legitimately condemned for desecrating the grave of the king of Edom because its actions broke the bond of friendship formed under David.

After the division of the kingdom, weak rulers on the Davidic throne were unable to maintain control of the area. These formerly interconnected nations reasserted their independence and regained their freedom. Amos viewed their moves to independence as a revolt, a breach of their contracts; they had rebelled against the Davidic ruler and the God who placed him on the throne. To make matters worse, they had even committed atrocities against former members of the association. Mauchline writes:

> The conclusion seems inevitable that Amos believed that these peoples had a bond of association which they had treated with contempt, a bond in the name of Yahweh which meant that, in spite of their political separation from Israel now, they were still a spiritual unity.[33]

This approach to Amos' oracles against foreign nations offers an interesting insight into two related passages in the Book of Amos.[34] We have examined the acceptance by many scholars of 9:7 as a statement that Yahweh rules over all nations. This may no longer be the most likely interpretation of the verse. A form-critical approach to the passage indicates 9:1–4, 7–8a constitute a single unit.[35] Beginning with a vision of the Lord sending an earthquake that shatters a cult sanctuary, possibly Bethel, the passage shows God as a divine warrior destroying the site and punishing the people. No matter where they attempt to hide from God the Avenger, they are sought out and annihilated. Amos 9:4b reads:

> And I (God) will set my eyes upon them
> for evil and not for good.

Verse 9:8a, employing the catchword "eyes," continues:

> Behold, the eyes of the LORD God are
> upon the sinful kingdom,

and I will destroy it from the surface of
the ground.

We are clearly dealing with a doom oracle. Located in the seam
between these two verses that refer to God's watchful eyes is a divine
oracle (v. 7) declaring that God's attitude toward the Ethiopians is
the same as his attitude toward the people of Israel, and that God
brought the Philistines from Caphtor, the Syrians from Kir, and the
Israelites from Egypt. The shift from direct address ("Are you not
like the Ethiopians to me," v. 7a) to third person ("Did I not bring
up Israel from the land of Egypt," v. 7b) produces an effective dis-
putational style. While v. 7 disrupts the unity of 9:4b and 8a, it clearly
specifies the object of Yahweh's wrath. But why is Israel compared
with these three peoples?

According to biblical traditions, the Philistines and the Syrians
had come into Canaan from the outside; the Philistines from Caphtor,
the Hebrew name for Crete (Deut. 2:23; I Sam. 30:14; II Sam. 8:18;
Jer. 47:4; Ezek. 25:16; Zeph. 2:5), and the Syrians from the Kir
region of Upper Mesopotamia (II Kings 16:9).[36] It is God, then, who
has given the Davidic empire its form. He brought three nations from
the outside; the rest were already there. During the settlement period,
God, as divine warrior, destroyed the Amorites (Amos 2:9)[37] and
gave Israel the land. Under David he established a united empire.
Because these foreign nations violated the covenant of brotherhood
formed through David, God would once again act as the divine warrior and would either destroy or exile them. Appopriately, God would
punish Syria by returning it to Kir from whence he had brought it
(1:5).

What about the future of Israel? The Davidic empire had begun
to dissolve under Solomon with the revolts of the Edomite Hadad
and of Rezon of Syria (I Kings 11:14–25). However, the rupture in
the relationship of Israel with the legitimate Davidic ruler did the
most damage. Israel has no excuse because it is an integral part of
the covenant people (Amos 3:2); above all, it should have known
better. Consequently, Israel is now compared with the Ethiopians,
the remotest people known to Amos.[38] Israel is declared a foreign
nation and will receive a judgment comparable to those mentioned
in Amos' oracles against foreign nations. The plain meaning of the
text, within this judgment context, is that just as God has no interest
in the distant Ethiopians, whom he did not bring into the land to

help form the Davidic empire, so he no longer has any concern for Israel! The text does *not* declare that god rules all nations. It *does* declare that God created the Davidic empire and that he is about to destroy nations within that empire that have rebelled against it.

Rebellion was the chief sin condemned by Amos; this is evidenced in his frequent use of the term *peša'*. Amos employs *peša'* to designate the crimes both of the foreign nations and of Israel. Its basic meaning is "rebellion" or "revolt," but it is often translated by the English word "transgression" or "crime." Preeminently a political term, *peša'* indicates a defiance of authority.[39] The deuteronomistic historians used the term in recounting the division of the united kingdom:

> So Israel has been in *rebellion* against the house of David to this day.
> (I Kings 12:19)

It was used again in describing an Edomite revolution against King Joram of Judah:

> In his days Edom *revolted* from the rule of Judah, and set up a king of their own. Then Joram passed over to Zair with all his chariots, and rose by night, and he and his chariot commanders smote the Edomites who had surrounded him; but his army fled home. So Edom *revolted* from the rule of Judah to this day.   (II Kings 8:20–22)

When used theologically, *peša'* designates sins against individuals (Gen. 31:36; 50:17; II Sam. 24:12 [H 24:11]; Prov. 28:24; 29:6), against nations (Prov. 28:2), and against God (Isa. 58:1; 59:12; Micah 1:5, 13; 3:8).

Many scholars have perceived that *peša'* in the Book of Amos referred to the act of afflicting the poor by those with political power and economic wealth. Possibly Amos had in mind a broader meaning of the term as well: the rebellion of Israel against Judah that resulted in the division of the kingdom and in the gradual dissolution of the Davidic empire. By revolting against the Davidic ruler, these nations, including Israel, had rejected the authority of Yahweh. This defiance of divine authority, manifest in the Davidic ruler, was graphically reflected in the atrocities foreign nations inflicted on brother nations of the empire and in the oppression of the poor within Israel.[40] Justice and righteousness would be achieved only when the empire was re-constituted and all its citizens live under the rule of Yahweh's true representative, the Davidic king. Throughout the Ancient Near East, moreover, one of the primary functions of kingship was assuring that justice prevailed in the land. Thus the Book of Amos contains no condemnation of Judah because Amos believed he was called primarily to condemn the northern tribes' revolt against Judah. That revolt had resulted in the formation of an illegitimate monarchy that failed to establish justice in the land. Judah had at least maintained legitimate Davidic rulers, although few of them had achieved the ideal of establishing justice and righteousness. The situation in Judah was therefore not as desperate as in the north. It was against the north that Amos must prophesy.

This broader approach sheds light on a second passage, 9:8b–12.[41] Most biblical scholars maintain this passage is an exilic or postexilic hope addition.[42] Four reasons have usually been offered to support that late date:

1. The abrupt change within 9:8 betrays the hand of a redactor. God has just declared that he will destroy Israel from the surface of the ground when he proceeds to say, "except that I will not utterly destroy the house of Jacob." This appears to be an addition referring to those who preserved the faith in the Babylonian exile (587–538).
2. Amos 9:9–10 uses sieve imagery to separate good persons from evil persons. Only the worthless pebbles, the sinners, will remain in the sieve and be destroyed. This implies that the grain, those who are righteous, will return to the land. Again, the passage appears to reflect the exilic period when sharp distinctions begin between the righteous and the wicked (Ezek. 18), the former looking forward to a restored Israel.
3. Amos 9:11 refers to a raising up of the booth of David. This is usually interpreted as a restoration of the Davidic line after its destruction in 587 by the Babylonians.
4. The special emphasis placed on Edom in 9:12 appears to fit best into the exilic period when Edom attacked a restored but still weak Judah.

While these arguments are impressive, there is substantial evidence supporting the authenticity of this passage within the earlier time frame. In the context of Amos' hopes for a reconstituted Davidic empire, the verse presents fresh insights into the material.

The use of the term Jacob in referring to Israel occurs six times in the Book of Amos: twice as simply "Jacob" (7:2, 5), twice as "the pride of Jacob" (6:8; 8:2), and twice as "the house of Jacob" (3:13; 9:8). These forms of the term convey Israel's election, which Amos insisted would not prevent Israel's destruction. Although Amos used the term "Jacob" twice when interceding for Israel (7:2, 5), perhaps to remind God of Israel's election, the record shows that God abhors "the pride of Jacob" (6:8) and will surely destroy that nation (8:7). The northern kingdom, therefore, should not believe its election has placed it in a favored position. Most scholars consider the first five references to Jacob in the Book of Amos as the authentic words of the prophet. That judgment further supports the position that the sixth reference to the preservation of "the house of Jacob" in 9:8 is also in the style of Amos.

"The house of Jacob" in 9:8b, contrasted with the "sinful kingdom" of Israel in 9:8a, has commonly been identified with Judah during the exile. If such an identification is correct, that would be the only place in Amos in which "Jacob" is so used. H. W. Wolff believes the "sinful kingdom" might mean "sinful kingship," in keeping with the usage of the Amos' school, which he dates ca. 760–730.[43] His interpretation of the text is that while the king will be punished the whole nation will not be destroyed. If 9:8b, however, is accepted as authentic Amos, the prophet has more likely declared that although Israel will be destroyed, he hopes a remnant of faithful Israelites will survive. Thus the phrase "house of Jacob" is now being used in the positive sense of God's continued concern for those who have not rebelled against him. The next four verses indicate what will happen to them.

In 9:9–10 the nation Israel has been defeated and exiled among the nations. But only those who are evil will be destroyed. Using the imagery of a sieve, Amos has likened the sinners in Israel to the pebbles that remain in the sieve and are thrown away. The text implies that the good grain is preserved.

Amos 9:11, with its reference to "the booth of David (*sukkat dāwîd*)" that is rebuilt, has always proved difficult to interpret. Booths are flimsy, wooden structures. The phrases "repair its breaches," "raise up its ruins," and "rebuild it as in the days of old" imply a more substantial structure. Some scholars contend the reference is to "the city of David" (Jerusalem),[44] but no textual support exists for this position. Suggestions have been limited only by the imaginations of the interpreters. J. L. Mays maintains the booth represents "the umbrella of David's rule"[45]; E. Hammershaimb claims it is "the Davidic dynasty"[46]; Wolff offers "Davidic imperium"[47]; W. R. Harper contrasts the restoration of "the hut of David" in its simplicity with the grand palaces and forts of Jeroboam II [48]; R. S. Cripps suggests "the ultimate reign of an ideal dynasty."[49]

In exploring the use of the Hebrew word *sukkâ*, John Mauchline has offered an intriguing interpretation of 9:11. He notes that Isaiah 4:5–6, using the imagery of the wilderness wanderings, refers to God's glory manifest in a cloud by day and by night smoke and fire placed over Zion. As God's glory appeared at Mount Sinai, it now appears at Mount Zion. That glory will provide a canopy (*ḥuppâ*) and a pavilion (*sukkâ*) to protect the people from harm. The term *ḥuppâ*, a parallel construction with *sukkâ*, means a bridal canopy. Mauchline believes this is royal imagery and refers to Yahweh's presence in the city as a true king who has wed Israel, his bride. Concerning the light this sheds on the Amos reference, he writes:

> that the words *the tent (sukkāh) of David* in Amos ix 11 were deliberately chosen and that the reference intended in that verse was to the restoration, not of the ruined city of Jerusalem but of the royal house of David, which had suffered an eclipse of its power at the end of Solomon's reign, and of David's kingdom which had been broken into fragments at the same time.[50]

Mauchline does not believe the text is a postexilic addition, although his interpretation of *sukkâ* could just as well fit that later period. To "repair its breaches," "raise up its ruins," and "rebuild it as in the days of old," according to Mauchline, refers to the restoration of the Davidic empire to its former glory before the division of the kingdom in 922, not before the destruction of Judah in 587.

A better interpretation of "the booth of David" has been offered by H. N. Richardson. He argues that the Hebrew word *sukkat* in 9:11 does not refer to a booth or a tent, but to the city of Succoth, an Israelite military garrison in the Transjordan region.[51] Richardson bases much of his theory on the work of Yigael Yadin, an Israeli archaeologist and militarist, who made the art of warfare in the Ancient Near East his specialty. In analyzing David's formation of an empire, Yadin has concluded that the site of Succoth on the Transjordan region figured prominently in the defeat of both the Ammonites and the Syrians.[52] The biblical account of the Ammonite defeat concentrates primarily on the David–Bathsheba–Uriah story (II Sam. 11:1–27). In describing the course of the war, however, we read that while Joab besieged the capital city of Rabbath–Ammon, the sacred ark accompanied the army into battle. Uriah's words to David are strange:

> The ark and Israel and Judah dwell in booths (*sukkôt*); and my lord Joab and the servants of my lord are camping in the open field.
>
> (II Sam. 11:11a)

Both military positions are strange. While the major professional army encamped in the open field, the ark and the reserves enjoyed the shelter of booths.[53] Yadin believes this campaign against Rabbath–Ammon makes better strategic military sense if the reserves and the ark were stationed at the city of Succoth. The chapter in II Samuel preceding this account describes Joab's first attack on Rabbath–Ammon from the south where Joab had to fight Ammonites in the front and Syrians to the rear. The battle had ended in a stalemate. For his next attack, Joab had taken a northern route, stationed his auxiliary troops at Succoth, and attacked Rabbath–Ammon from the direction of Syria. When the city was about to fall, Joab sent to David to "gather the rest of the people together, and encamp against the city, and take it" (II Sam. 12:28). Because the "rest of the people" denoted the militia stationed at Succoth, that city assumed a major military importance in David's extension of his kingdom into the Transjordan region.

The term *sukkôt* appears again in I Kings 20:12, 16, in the context of a conflict between the king of Syria Ben-hadad and the king of Israel Ahab. Yadin argues that the sending of messages and the positioning of the armies favor interpreting *sukkôt* as a place-name rather than booths. That place-name then becomes the Transjordan city of Succoth that lay on the route to Samaria, an ideal position from which Ben-hadad attacked Ahab.[54]

In the last half of the ninth century, this Transjordan area had been devastated by the Assyrians under Shalmaneser III. When Shalmaneser withdrew because of problems to the north of his kingdom, Hazael of Syria gained control of both Cisjordan Israel and the Transjordan region. Perhaps not until the time of Jeroboam II (786–746) had this territory been recovered by Israel (II Kings 14:25). There is no indication in the text that Jeroboam II had any intention of rebuilding Succoth. It undoubtedly lay in ruins and became for Amos a symbol of a reconstituted Davidic empire. Commenting on Amos' oracle of hope in 9:8b–12, Richardson writes:

> In this he looked forward to the reunification of the two kingdoms under the Davidic dynasty. And a symbol of that reunification was the rebuilding of David's Succoth as a prelude to the re-establishment of the glorious rule of Yahweh's anointed from Elath to the Euphrates.[55]

Verse 9:12 precisely describes what the reestablishment of David's empire would be like:

> that they may possess the remnant of Edom
> and all the nations over whom my name is called.[56]

The phrase "All the nations over whom my name is called" refers to those nations addressed in the oracles against foreign nations (1:3–2:3). The Davidic empire would be reconstituted; Israel and Judah would be reunited and occupy all those lands that once made up the original Davidic kingdom.

Edom has been singled out as possessing land that the united kingdom will occupy because Edom had violated the close bond of fellowship it had formed with the Hebrew nation. The tradition concerning Esau (Edom) and Jacob (Israel) demonstrates how their original close kinship changed into animosity. Recurrent conflict characterized the relationship. Perhaps Israelites were those people sold to Edom as slaves by the Philistines (1:6) and by the city of Tyre (1:9). According to II Kings 8:20–22, Jehoram of Judah (849–843) lost control of Edom; Amaziah (800–783) defeated Edom and extended his control to Elath (II Kings 14:7, 10, 22); Edom again broke free in 735 (II Kings 16:6) at the beginning of the reign of Ahaz. Amos' ministry may have continued in the south until 735, in which case this reference to Edom (Amos 9:12) could be related to the threat to the Davidic line when Ahaz was attacked both from the south (Edom) and from the north (Syro-Ephraimite alliance).[57]

We have devoted considerable attention to the place of the Davidic empire within the thought of Amos. Whether or not the theory is valid depends on how well it helps us interpret Amos' message, especially his oracles against foreign nations. We now turn to this task.

Amos' Oracles against Foreign Nations

Damascus

Each of the oracles against foreign nations begins with the same phrase:

> For three crimes (name),
> and for four, I will not cause it to return.
>
> (author's translation)

We have already had occasion to note the sapiential style of the graduated numbers.[58] How to translate the clause that follows the numbers is difficult to determine. Scholars differ concerning the reference of the objective pronominal suffix attached to the verb. In the RSV it refers to the punishment: "I will not revoke the punishment." Similarly, the NEB translates "I will grant them no reprieve." If this were the case, however, a feminine not a masculine pronoun would have been used, referring to the feminine noun "fire" as the punishment.[59] It could allude to the Assyrians as the agents of punishment, although the Book of Amos contains no explicit reference to that nation.[60] The reference probably does not apply to the condemned nation, indicating that nation will not return after its deportation. The latter is a possible interpretation for the Damascus, Philistine, Ammon, and Moab oracles, but not for the Tyre and Edom oracles. A literal translation most likely suggests that the "it" refers to the word of God. In the first two visions (7:1–3, 4–6), Amos interceded for Israel and the visions ceased. The next two visions (7:7–8; 8:1–2) close with the words "I [God] will never again pass by them." Amos did not intercede for Israel; God did not cause the vision to return to him unfulfilled. The vision completed itself and judgment fell upon the nation.[61] As in all the oracles against foreign nations in the Book of Amos, judgment comes through the mythological symbol of divine fire that devours the enemy (1:4, 7, 10, 12, 14;2:2).[62]

Damascus, the ancient capital of Syria and the largest and most influential city in the Aramaean states, is northeast of Israel. Damascus is accused of treating harshly during wartime the Transjordan region of Israel known as Gilead. "Threshing Gilead with threshing sledges of iron" may either be a metaphor for harsh treatment or a literal description of how enemy people were tortured, having threshing instruments with flint or iron teeth drawn across them. In any case, the conduct is barbaric enough to be condemned.

The precise history of Israel's conflict with Syria is difficult to determine, because legendary stories of Elijah and Elisha have caused a dislocation in the sequence of events. Nevertheless, between Syria and Israel in the Gilead region there are four main periods of warfare that could provide the occasions to which Amos' oracle refer. Because this material is also relevant for the oracle against Ammon, the four periods are detailed here.

The first period of conflict began during the reign of Solomon when Rezon, son of Eliada, led a revolt and freed Damascus from Israelite domination (I Kings 11:23–24). The text continues:

> He was an adversary of Israel all the days of Solomon, doing mischief as Hadad did; and he abhorred Israel, and reigned over Syria.
>
> (I Kings 11:25)

The conflict continued during the reign of Baasha of Israel (900–877), when King Asa of Judah (913–877) paid the Syrian king to break his treaty with Israel and harass her northern cities (I Kings 15:18–22). Syria and Israel more than likely warred on the battlefields of Gilead during this period.

The second possible period of conflict occurs during the reigns of Ahab of Israel (869–850) and Jehoshaphat of Judah (873–849). Recent studies of I Kings 20 and 22, however, raise doubts that Ahab even engaged in warfare with Syria.[63] Apparently throughout Ahab's reign an alliance prevailed between Ahab and Hadadezer[64] of Syria to prevent Assyrian aggression in the area. The conflict between Israel and Syria, which the deuteronomistic historians placed at this time, better fits a later period. This reconstruction also clarifies Syrian history: no longer are there three Ben-hadads,[65] but simply Hadadezer, Hazael, and Benhadad.

A third major period of conflict between the two countries begins during the reign of Hazael (II Kings 8:28) with contention over the Gilead region:

> In those days the LORD began to cut off part of Israel. Hazael defeated them throughout the territory of Israel, from the Jordan eastward, all the land of Gilead, the Gadites, and the Reubenites, and the Man-

assites, from Aroer, which is by the valley of Arnon, that is, Gilead
and Bashan.   (II Kings 10:32–33)

Although the history of these wars between Syria and Israel cannot be
accurately determined, one possible and commendable reconstruction
suggests the reign of Jehoahaz (815–802) as the beginning of Israel's attempt to free itself from the Syrian domination of Hazael and his successor Ben-hadad (II Kings 13:4–5). Jehoahaz had been only partially
successful, and Syria's counterattack had resulted in a siege of Samaria. However, the westward campaign of the Assyrian king Adad-
Nirari III so weakened Syrian resolve that Jehoahaz's successor
Jehoash (802–786) was able to defeat Ben-hadad on three separate
occasions and to regain Israel's independence (II Kings 13:14–25).[66]

A fourth period of conflict between Syria and Israel dates from
the latter part of the reign of Jeroboam II (786–746).[67] Following the
death of Adad-Nirari III in 782, the kingdom of Urartu had harassed
Assyria on its northern border and had prevented it from dominating
Syria and Israel. Not until the reign of Tiglath-pileser III in 735 did
Assyria reassert control over this region of the fertile crescent. Between ca. 773 and 735, therefore, Syria and Israel were free to renew
their ancient conflicts. Jeroboam was able to extend his kingdom from
the entrance of Hamath in the north as far as the Dead Sea in the
south (II Kings 14:25; Amos 6:14) and even to secure control of
Damascus (II Kings 14:28). If this fourth period does encompass the
specific conflict mentioned in Amos, then it would have the added
advantage of making Amos 6:13 a possible reference to the victories
of Jeroboam II over the two Syrian towns of Lo-debar and Karnaim.[68]

No one knows for certain what specific events Amos had in mind
when he condemned Damascus for its harsh treatment of Gilead. Of
the four suggested periods, the evidence supports either the contemporary period of Jeroboam II or the period of Jehoahaz–Jehoash. No
clear evidential choice can be made between them.

Philistia

The Philistines resided along the coast of the Mediterranean Sea,
southwest of Israel. They controlled six major cities in the region,
four of which are mentioned by Amos. Scholars have noted the ab-
sence of Gath in the list of Philistine cities condemned by Amos.
Some have said this presupposes Gath's destruction by Sargon II of

Assyria in 711, which would make this oracle a later addition. However, Gath might still have been in ruins from Hazael's attack (II
Kings 12:17 [H 12:18]). Or Gath might not have been independent
at this time, but rather under the control either of Judah (II Chron.
26:6) or of Ashdod.[69] There is no compelling reason, furthermore,
for Amos to have mentioned Gath in this context. From cities mentioned in other oracles, there is no reason to expect an inclusive list
here. Few scholars regard the omission of Gath as serious evidence
against the authenticity of the oracle.

The Philistines, according to the accusation, dealt in slave trade,
probably for economic gain. The Philistines sold the entire population
of some communities to the Edomites.[70] Although the text neglects
to identify those who were enslaved, the most likely people geographically would have been either Israelites or Judahites. Possibly
Amos was recalling the occasion when the Philistines carried away
the sons and wives of Jehoram of Judah (II Chron. 21:16–17).

The punishment for Philistia was the same as for Damascus.[71]
Those who sit upon the thrones at Ashdod and Ashkelon will be
destroyed.

Tyre

Tyre was a city-state on an island off the coast of Phoenicia, northwest
of Israel. Alexander the Great altered the geography of the region
by constructing a land bridge from the mainland to the island in order
to capture Tyre. Thereafter, Tyre formed a peninsula.

The accusation against Tyre, like Philistia, is engaging in slave
trade. The repetition of the indictment has led some scholars to regard
the oracle as a postexilic addition to the series.[72] They argue that the
author purposely avoided the word *gālût* ("exile"), because it had
become a technical term to designate the Babylonian exile. Furthermore, the use of *zākar* ("to remember") in Amos 1:9, first found in
Deuteronomy, Ezekiel, and Deutero–Isaiah, would indicate late
usage.[73] Others have not found these arguments persuasive,[74] because
the linking of Philistia and Tyre was not uncommon (see Joel 4:4–8;
Jer. 47:4; Ezek. 5:16–17; Zech. 9:3–6; Ps. 83:5–8).

It is impossible to determine with any accuracy to what Amos'
indictment was specifically referring. Most of the attention has
been given to "the covenant of brotherhood" that Tyre broke by
engaging in slavery. The accusation may refer to the Hiram of

Tyre agreement to help David build his palace (II Sam. 5:11).[75] But more likely Amos' reference is to the same Hiram's treaty with Solomon (I Kings 5:1–18 [H 5:15–32]; 9:13).[76] The account has a remarkable beginning:

> Now Hiram king of Tyre sent his servants to Solomon, when he heard that they had anointed him king in place of his father; for Hiram always loved David.   (I Kings 5:1 [H 5:15])

Concerning Hiram's disappointment when Solomon attempted to settle his debt with the king of Tyre by offering him twenty Galilean cities, Hiram replied:

> What kind of cities are these which you have given me, my brother?
> (I Kings 9:13)

That Hiram "loved" David and referred to Solomon as "my brother" indicates the "covenant of brotherhood" binding the two nations.[77] If the people Tyre had sold into slavery to Edom were Israelites or Judahites, then the covenant agreement had been broken and Tyre was justly condemned.

Given the religiously corrupting nature of Ahab's marriage to Jezebel, this union would hardly be the "covenant of brotherhood" Amos cited. Also possible, but not probable, is the supposition that it was the Edomites who had failed to remember "the covenant of brotherhood."

No specific judgment was pronounced on Tyre. Using holy-war imagery, Amos declared that fire would devour the city.[78]

Edom

Edom is a region southeast of Israel, directly south of Moab, and east of the deep fault (the Arabah) south of the Dead Sea. Edom had been included as a vassal state when David formed the empire. As we have seen, Edom had begun to break free of Israelite control during the reign of Jehoram of Judah.

The oracle against Edom is often considered a postexilic addition to the series.[79] Unquestionably, the brother pursued with the sword is Israel. The issue is into which historical period the oracle best fits. After the fall of Jerusalem to the Babylonians, Edom had conducted military raids against her northern neighbor, seizing border territories and treating refugees ruthlessly. A body of anti-Edomite literature had arisen of which Obadiah and Lamentations 4:21–22 are the clearest examples.[80] Another argument concerns style. Does the style of the oracle reflect the work of a redactor? As with the oracles against Tyre and Judah, the punishment, the standard form of fire that devours, lacks specific details found in the genuine oracles; also lacking is the closing refrain "says the LORD."

A number of scholars, however, are not convinced by these arguments.[81] Michael Fishbane of Brandeis University, in examining the Hebrew words translated "brother" and "pity" in Amos 1:11, finds an interconnectedness in the treaties of the Ancient Near East. In those treaties the Akkadian cognate for "brother" is the technical term for "treaty partner," while the cognate for "pity" is a verb best translated "to love, to recognize."[82] As Fishbane cites, these are the words used on a personal level in the biblical material when Jonathan called David his "brother" (I Sam. 20:29) and made a covenant with him because he "loved" him:

> When he had finished speaking to Saul, the soul of Jonathan was knit to the soul of David, and Jonathan loved him as his own soul. . . . Then Jonathan made a covenant with David, because he loved him as his own soul.   (I Sam. 18:1, 3)

Not without reason, therefore, is Edom called a "brother" in diplomatic relations with Israel (Num. 20:14; Deut. 2:4, 8; Obad. 10) and in patriarchal narratives. This, Fishbane argues, is language of the covenant-treaty that has its origin in David's inclusion of Edom as a vassal state in his empire. Fishbane concludes that the most likely historical setting for Amos 1:11 is Edom's revolt during the reign of Jehoram (II Kings 8:20–22). In Fishbane's view, the Edom passage is a genuine Amos oracle.

Keith N. Schoville has treated the first three nations of Philistia, Tyre, and Edom as a unity.[83] He argues that all three had formed a "covenant of brotherhood" with David. Before David became king, a close relationship existed between Achish of Gath and David (I Sam. 27–29). In addition, both David and Solomon formed an alliance with Hiram of Tyre. Edom's close kinship with Israel has been documented in the preceding paragraph. Schoville suggests that Jehu's submission to Shalmaneser III in 841 dissolved the alliance of these four states and led Philistia, Tyre, and Edom to punish Israel by enslaving part of its population.

In describing the punishment by destructive fire that will descend upon Edom, Amos mentioned by name two places: Teman, the largest city in the southern section of the land; and Bozrah, a major fortress in the north. In this way Amos indicated the total devastation of the nation for its revolt against Judah.

Ammon

The kingdom of Ammon lay in the Transjordan region, north of Moab. Through military conquest, David incorporated it within his kingdom. But soon after Solomon's death, it regained independence. Amos, however, condemned Ammon for breaking a boundary treaty with Israel in an attempt to seize Gilead. Furthermore, the accusation was for wartime atrocities that violated human rights. Although these atrocities might have occurred on a number of occasions historically, perhaps the most likely was the period of Syrian aggression under Hazael. Such conflict was frequent; in Amos' own time, Jeroboam II had resorted to warfare to reestablish his kingdom in this region (II Kings 14:28).

Even more than in the other oracles against foreign nations, this oracle contains holy-war imagery to describe God's judgment on the land. Especially are the rulers of Ammon—the king and princes in the capital city of Rabbah—singled out for punishment.

Moab

Moab was a nation east of the southern half of the Dead Sea, between Edom and Ammon. David had made it a vassal state in his empire; when the kingdom was divided it passed under the control of Israel. The famous Moabite stone, erected by King Mesha of Moab, celebrated that nation's liberation from Israelite domination during either the reign of Ahab or Jehoram.[84] Possibly Jeroboam II retook the area, but the evidence is too meager to be certain (II Kings 14:25). The Moabite stone also indicates that Kerioth, mentioned in Amos 2:2, contained a sanctuary of the Moabite god Chemosh.

Moab's crime was the desecration of the grave of an Edomite king, possibly by using his bones to make a lime whitewash for use on Moabite homes. Whatever precisely happened, the crime was treating an enemy in a barbaric manner. The basis of Yahweh's concern is twofold. First, because it had once been part of the Davidic empire, Moab should have treated associate members humanely. Second, Moab's atrocity had been directed against Edom, a people who were in a special covenant of brotherhood with the Hebrew people.

In addition to punishing Moab with the fire that devours, Amos' oracle contains such holy-war imagery as "uproar," "shouting," and "sound of trumpet." For their crimes, the ruler and the princes of the land will be slain.

We have seen that the reestablishment of the Davidic empire was a genuine hope of the prophet Amos. The grounds for his condemnation of neighboring nations were their violation of the covenant association with David on which the empire was based. The form of the reunited empire was not made clear by Amos. Perhaps the neighboring nations would once again accept the rule of David's line. But references to exile (1:5, 15), to destruction (1:5, 8, 10, 12, 14; 2:2), and to the possession of the remains of Edom (9:12) support the view that Amos thought these lands would become occupied by a united Israel.

Once the empire had been reconstituted, Amos believed worship acceptable to Yahweh would be reestablished and justice would prevail in courts supervised by Davidic rulers. The next two chapters will explore these subjects.

5

Amos' Polemics against Northern Shrines

AMOS WAS a Judahite called to prophesy to Israel. In obedience to that call, he journeyed north and prophesied at Bethel (7:10–17) and probably at Samaria (4:1–3; 6:1–14). In the previous chapter, we saw that the prophet's message condemned the surrounding nations for violating the covenant of association, the foundation of the Davidic empire. However, the major object of Amos' scathing message of God's judgment was Israel itself. What was there about that kingdom that Amos believed deserved God's wrath? The answer frequently given is that Amos condemned the north for its oppression of the poor. One may argue that for Amos the essence of religion was the call for social justice. Yet much of the Book of Amos also attacks cultic religion. This should be expected, for in the Ancient Near East the cult, especially the royal cult, was responsible for instituting and maintaining social justice.

Religion in the Northern Kingdom of Israel

The deuteronomistic historians never tire of condemning Israel for Baal worship. For them, apostasy began with the first king of Israel, Jeroboam I, who erected two golden calves at Bethel and Dan, established high places, instituted a new line of non-Levitical priests, and introduced a new festival calendar (I Kings 12:25–33). These historians further contended that Baal was worshiped in the north in the form of golden calves and through sacred prostitution rites. They therefore condemned the northern kings for walking in the way of Jeroboam I.[1] The deuteronomistic historians did not, however, condemn the use of Canaanite furnishings found in Solomon's temple in Jerusalem. There was no criticism of that temple's basic structure, of the twelve bull images to support the bronze sea, of the elevated altar for burnt offerings, or of the winged cherubim—all, to the best current scholarship, based on Canaanite models. For these historians, the only proper place to worship Yahweh was in Jerusalem at Solomon's temple.

The deuteronomistic historians give us a biased interpretation of the history of Israel. Writing after the Josiah reforms of 621 B.C.E., they reflect that king's attempt to purify the faith by tearing down competing sacred places and centralizing worship in Jerusalem (II Kings 22–23). We have seen, however, that a more balanced evaluation of religion under Jeroboam I recognizes his role as a Yahweh worshiper attempting to establish a rival cult to Jerusalem, a cult loyal to the northern monarchy. There were Canaanite influences in the northern religion, but this was also the condition of religious practice in the south. There is no support for the position that northern religion under Jeroboam I was more Baalistic than southern religion under Rehoboam, Solomon's son. Even the prejudiced deuteronomistic historians admitted that Rehoboam built high places, erected Canaanite cult objects such as pillars and Asherim, and introduced male cult prostitutes into Judah (I Kings 14:21–24). We now know that religious syncretism was occurring throughout the periods of the settlement, the united monarchy, and the early years of the divided kingdom. Biblical scholars once held that there was a cultural conflict between Israelite pastoralists and Canaanite agriculturalists. Apparently even in this area there was little distinction between them.

Not until the Omride dynasty did a sharp religious conflict occur between Yahwism and Baalism. The two protagonists were Jezebel and Elijah. Jezebel, daughter of a Sidonian king and wife of King Ahab of Israel, introduced Tyrian Baalism into Israel and favored its position in Samaria (I Kings 16:29–33). Elijah, the champion of Yahweh, led the opposition. This religious conflict eventually resulted in a political revolution and the establishment of the Jehu dynasty. But neither Elijah nor Jehu, with all their opposition to Baalism, ever opposed the golden calves at Bethel and Dan. Apparently they considered as legitimate Yahwism the form of worship at those sites.[2]

The effectiveness of this religious purge of Baalism during the Jehu revolution is unknown.

The next major period of religious reform arose during the reigns of Hezekiah (715–687) and Josiah (640–609) of Judah. Both kings razed sanctuaries at high places outside Jerusalem. It was Josiah, however, who was most successful; his reforms led to the destruction in 621 of the northern cult center at Bethel. The religious rites performed at Bethel during that time are difficult to determine. The northern kingdom had been destroyed in 722/1; it was no longer the "king's sanctuary" (Amos 7:13). According to the deuteronomistic historians, when the Assyrians deported the prominent citizens of Israel and repopulated the region with political exiles from other areas of the Fertile Crescent, they also imported the gods of those exiles (II Kings 17:29–31). This may well have occurred. Because foreign cults had a prominent place in Jerusalem even after the reforms of Josiah (see Jer. 7:16–31; Ezek. 8:1–18), one may reasonably argue their prevalence in the northern region after the Assyrian conquest. Morton Cogan's study of Assyrian imperialism and religion reveals that vassal states bore no cultic obligations to Assyria. It was only in territories officially annexed as provinces by Assyria, like Israel after 722/1, that an Assyrian cult was introduced. Native cults remained intact, although the god Ashur was probably recognized as head of the pantheon.[3]

The primary sources of information about religion in Israel during the eighth century are the prophetic books of Amos and Hosea. The Book of Hosea is filled with anti-Baalistic language.[4] Hosea believed his people were unfaithful to Yahweh and went "awhoring" after Baal. They participated in fertility cult life in order to assure both the productivity of the land and the reproduction of children. God's judgment on them would be famine on the land and sterile wombs (Hos. 9:11–14). Even Hosea's life with the unfaithful Gomer symbolized the waywardness of the northern peoples who loved Baal more than Yahweh (2:1–23).

In the Book of Amos, however, the situation is quite different. Amos' attack on northern shrines shows he approached the subject from a different perspective than did Hosea. *There is not one overt reference to Baal in the Book of Amos.* The symbolism of fertility worship is largely missing. The explanation for this difference between the attitudes of Amos and Hosea toward cultic life in the north reveals much about Amos' view of the Davidic empire.

Amos' Attitude toward the Northern Cult

In *Amos among the Prophets*, Robert B. Coote argues that the A stage of the Book of Amos consists of oracles against the upper classes within the northern kingdom for oppressing the poor and needy. The B stage, attributed in its present form to the Bethel editor writing at the time of Josiah's reforms, contains the anticultic material. The C stage, composed during the exile, contains hope material assuring the community that God will return them to the land. The present book, therefore, consists of at least three identifiable layers of material. Coote is precise in separating what he designates the original words of Amos from those of the redactors. He considers only forty verses as authentic Amos, arguing that those texts attacking the northern cult reflect the period of Josiah's reforms.

The brevity of the Book of Amos makes it highly unlikely that literary critics have enough material to discover such detailed distinctions among the various stages of the text.[5] The claims are reminiscent of the excessive use of the documentary hypothesis by a previous generation of biblical scholars; one critic subdivided the Priestly document into seven major sources, each of which might have second and third sources, together with redactors and even second redactors (a mathematical possibility of eighty-four sources!).[6] The Pentateuch, however, because of its size and long history of composition, invites more obvious distinctions among the various stages of composition. The Book of Amos is a mere nine chapters in length with a much shorter history of composition. For Coote to define Amos' style from only forty verses is extremely hazardous.

If the anticultic passages in the Book of Amos are considered genuine Amos, there are at least four major approaches to understanding Amos' attitude toward cultic religion in Israel.[7]

Amos' Opposition to All Cultic Religion

Some scholars argue that Amos called for a pure form of spiritual religion that leads to the just treatment of the poor.[8] Such phrases as "I hate, I despise your feasts, and take no delight in your solemn assemblies" (5:21) and "do not seek Bethel, and do not enter into Gilgal or cross over to Beer-sheba" (5:5) are understood as a total rejection of all cultic life. All that Yahweh demands of his people is "to let justice roll down like waters, and righteousness like an everflowing stream" (5:24).

Amos' Opposition to the Substitution of Sacrifices for Social Justice

Given our knowledge of the Ancient Near East, it is unlikely that the Hebrew prophets rejected all cultic life.[9] If this were the case, they were the only persons in the ancient world who believed religion could be properly practiced without sacrifices. While cultic life is not limited to sacrifices, they are an essential part of religious practice in the Ancient Near East. These scholars argue that Amos did not object to sacrifices if they were accompanied by the just treatment of the poor and the needy. The people had allowed the cult to dominate their lives and had failed to care for the widow and the orphan. For this they stood condemned. An Amos-like verse from Isaiah summarizes the ideal relationship between cult and ethics: "I (God) cannot endure iniquity *and* solemn assembly" (1:13b). Cultic life apart from ethical conduct is abhorrent to God.

Amos' Criticism of Canaanite Cultic Influences

A considerable number of scholars hold that Amos' attack on cultic religion in the north went beyond merely striking a balance between social justice and sacrifice. Amos appears also to criticize Canaanite influences on the cultic religion of Israel. The debate centers on whether Amos condemned the north for worshiping Yahweh through Canaanite cultic practices or for worshiping foreign deities. The anticultic texts are open to variant interpretations.

We know from comparative studies that the Hebrew people borrowed extensively from the Canaanites when they settled in the land. In the area of religion, they took possession and rededicated to Yahweh old Canaanite sacred sites such as Shiloh, Bethel, Gilgal, Dan, Shechem, and Beer-sheba.[10] The form of their worship was heavily dependent on Canaanite practices. The priestly orders, the sacrificial methods, and the religious calendar originated in Canaanite culture.

Was Amos aware that the cult was overlaid with Canaanite practices? Amos 5:25 reads: "Did you bring to me sacrifices and offerings the forty years in the wilderness, O house of Israel?" The question begs for a negative answer. This verse, however, is a deuteronomistic addition to the text, embodying the tradition that the first generation of Israelites obeyed God's law without the demand for sacrifice.[11] The same tradition is found in Jeremiah 7:21–23. There is also evidence Hosea believed that life in the wilderness was an ideal period of faithfulness to Yahweh (see Hos. 9:10) and that Hebrew faith became corrupted upon entrance into the land. Hosea longed for the day Yahweh would return the people to the wilderness and rebetroth them to himself. Away from the land, removed from all Baal influences, they would learn to call God "my húsband" *('îšî)* instead of "my Baal" *(ba'lî).*[12]

Apparently Amos did not share this idealized interpretation of the wilderness wanderings, but he was aware that the cult had become so elaborate that social justice was ignored. The famous passage in 5:21–24 refers to God's rejection of feasts, solemn assemblies, burnt offerings, cereal offerings, peace offerings, and cultic music. These practices are continually referred to as "your" (the people's) expressions of cultic life, not Yahweh's. God takes no pleasure in an abundance of festivals and sacrifices, but in justice and righteousness.

The clearest references in the Book of Amos to Canaanite influences on Hebrew faith are in 2:7–8 and 6:4–7. Until recently these texts were interpreted as condemning the upper classes for gross immorality and luxurious living while the poor lacked the bare essentials of life.[13] However, in 1961, at Ras Shamra, an Ugaritic text was discovered that offers new insights into these passages. It describes in rather graphic detail a banquet called a *marzēaḥ*, given by El for the divine assembly. In this festival, the gods overeat and overdrink until they are satiated and intoxicated. But El outdoes them all: He becomes delirious and hallucinates a creature with horns and tail, and at the close of the banquet, collapses in his own excrement and urine. The text concludes with a few broken lines to suggest a proper medical treatment for those who suffer from alcoholic intoxication.[14]

M. H. Pope summarizes his findings:

From the various strands of information, we gather that the *marzēaḥ* was a social and religious institution which included families, owned property, homes for meetings and vineyards for wine supply, was associated with specific deities, and met periodically, perhaps monthly, to celebrate for several days at a stretch with food and drink and sometimes, if not regularly, with sexual orgies.[15]

The word *marzēaḥ* is used twice in Hebrew Scriptures. Jeremiah 16:5 associates it with mourning rites. Because of the frightful destruction at hand, the people will forego all mourning rites; they will not enter the house of *marzēaḥ*. There follows a list of mourning rites that include lamentations, lacerations, shaving the head, eating, and drinking. The second reference is in Amos 6:7 in which the word *mirzaḥ* refers to a revelry. Amos 6:4–7 is a woe oracle directed against those who engage in wild orgies that include stretching out upon couches, eating lambs and calves, singing songs, drinking excessively,[16] and anointing themselves with costly oil. There is little doubt that these customs relate to Canaanite practices.[17]

It may appear strange that the same word is associated both with mourning rites and revelry. But in fertility cults life and death are essential aspects of the natural order; death is overcome by renewal of life.[18] In contemporary culture, a New Orleans funeral takes the form of a jazz celebration. And in a quieter mood, the mourners at a funeral not uncommonly gather at the home of a deceased's close relative for food and drink provided by neighbors.

Did Amos condemn the northern people for worshiping a foreign deity through such rites? No foreign deity is mentioned in the text. More likely Amos condemned the people for worshiping Yahweh through a cultic celebration patterned after the Canaanite *marzēaḥ* festival.

Amos 2:6–8 is usually interpreted as an indictment of Israel for mistreatment of the poor and for gross sexual immorality. Hans Barstad, in *The Religious Polemics of Amos,* urges that while vv. 6–7a deal with oppression of the poor, vv. 7b–8 refer to the *marzēaḥ*.[19] Barstad's translation of vv. 7b–8 reads:

> A man and his father go to the maid,
> profaning my holy name,
> in front of every altar they lie down,
> on garments seized in pledge,
> and in the house of their gods they drink
> the wine given as rates.[20]

These verses have always been enigmatic. Does the reference to a man and his father going to the maid imply sexual immorality? Most critics believe it does.[21] But Barstad argues that "the maid" (*hanna'ărâ*) is neither a harlot nor a sacred prostitute, but a *marzēaḥ* hostess. "A man and his father" indicates the family nature of the rite. The profaning of God's holy name occurs when, in violation of

the law, clothing taken from the poor as collateral for a loan are not
returned at night (Exod. 22:26–27) but are used as garments upon
which to stretch out during the *marzēaḥ*. During the rites the cele-
brants consume wine acquired from fines probably exacted from the
poor, given the context. Most critics translate *'ĕlōhēyhem* "their
God," which implies Israel worshiped Yahweh but in a corrupt manner. The translation Barstad chooses is "their gods," an indication
that foreign deities, not Yahweh, were being worshiped through these
rites. The argument is impressive.

With less success, Barstad also maintains that Amos 4:1 refers to
the worship of foreign deities:

> Hear this word, you cows of Bashan
>   who are in the mountain of Samaria,
> who oppress the poor, and crush the needy,
>   who say to their husbands, "Bring,
>   that we may drink!"

The majority of commentators agree "the cows of Bashan" refer to
the wives of wealthy aristocrats in Israel.[22] Bashan, in the Transjordan
area, was famous for its fine pasturelands. The wives enjoyed luxurious living; they were "the pampered darlings of society in Israel's
royalist culture."[23] Barstad detects no less than four references to a
foreign cult in this one verse!

1. Just as the bull represents fertility in the Ancient Near East, so
   does the cow. This is true in Egypt and Mesopotamia, but especially in the Baal–Anath myth in Canaanite religion.
2. These "cows of Bashan" are "in the mountain of Samaria." Twice
   in the Psalter "Bashan" appears within a mythical context:

> Many bulls encompass me,
>   strong bulls of Bashan surround me;
> they open wide their mouths at me,
>   like a ravening and roaring lion.
>   (Ps. 22:12–13)

> O mighty mountain, mountain of Bashan;
>   O many-peaked mountain, mountain of Bashan!
> Why look with envy, O many-peaked mountain,

at the mount which God desired for his abode,
yea, where the LORD will dwell for ever?

(Ps. 68:15–16)

Behind these texts is undoubtedly an old mythic tradition that regarded the mountain of Bashan as a cosmic mountain, comparable to Saphon and Zion. In Amos, the cows associated with the religious tradition of Mount Bashan are now honored in Mount Samaria.

3. The text does not use the ordinary word for husband (*ba'al*), but rather *'ādôn*. Barstad argues that Amos purposely used this word because Canaanite religion frequently used the term to refer to a god. This provides additional support for interpreting the cows as female deities.

4. The "cows of Bashan," from the upper classes because they oppress the poor, ask their husbands to bring them drink. Barstad finds here an allusion to the *marzēaḥ*.

His argument is somewhat forced; he reads into the text more than is there. The majority of scholars still continue to regard the passage an attack on those in the upper classes who oppress the poor and the needy.

Three other verses contain references to foreign cults, according to Barstad. Amos 5:26 threatens that when the end comes, Israel will carry into exile the images of Sakkuth and Kaiwan, probably fixed atop standards.[24] Sakkuth and Kaiwan were Assyrian astral deities. These deities, however, were probably not introduced into Israel until the Assyrian occupation and resettlement of the area (II Kings 17:29–31). In any case, most scholars regard 5:25–27, which idealizes the wilderness wandering, as a deuteronomistic addition to the text.[25]

Amos 6:13 may be a play on words, referring to idols that are regarded as nothing (*Lŏ-dābār* = no thing) and idols that have two horns (*qarnāyim* = dual form of horns).[26] The context, however, supports interpreting these as the names of sites taken by Jeroboam II during a period of expansion.[27]

The third verse, 8:14, is often cited to support the thesis that Amos condemned the northern kingdom for worshiping foreign gods:

Those who swear by Ashimah of Samaria,
and say, "As thy god lives, O Dan,"

> and, "As the way of Beer-sheba lives,"
> they shall fall, and never rise again.

The verse refers to the religious practice of taking an oath in the name of God. Many references in Hebrew Scripture indicate that to swear by the name of God was expected of his people:

> You shall fear the Lord your God; you shall serve him, and swear by
> his name.   (Deut. 6:13)

Conversely, to swear by the name of another god was condemned:

> Therefore be very steadfast to keep and do all that is written in the
> book of the law of Moses, turning aside from it neither to the right
> hand nor to the left, that you may not be mixed with these nations
> left here among you, or make mention of the names of their gods, or
> swear by them, or serve them, or bow down yourselves to them.
>   (Josh. 23:6–7)

Amos 8:14 refers to swearing by "Ashimah of Samaria." According to II Kings 17:30, Ashima was a Syrian deity from Hamath who was not introduced into the northern kingdom until the period following the Assyrian destruction of Samaria. Is this historically accurate? Some scholars hold the text is unhistorical and postexilic,[28] while others maintain it contains details that are historically accurate.[29] The Masoretic text of Amos 8:14 does not read *'ašîmā;* it reads *'ašmat* (construct of *'ašmâ*), which may be translated "guilt." The "guilt of Samaria" may refer to the golden calf located in either the capital at Samaria or Bethel (Hos. 8:5).[30] Barstad argues that the Hebrew word *'ašmâ* appears only in postexilic texts; hence, it was substituted in Amos 8:14 for the name of the Syrian goddess.[31] But *'ašmâ* occurs four times in Leviticus (4:3; 6:5 [H 5:24], 7 [H 5:26]; 22:16). While the final form of the Book of Leviticus is undoubtedly postexilic, the work contains legal collections that probably represent the traditions of various Israelite sanctuaries during the monarchical period. Furthermore, the name of the Syrian goddess *'ašîmā* appears only in II Kings 17:30, a late preexilic or exilic text. After a lengthy discussion of the subject, Barstad concludes that Amos condemned Israel for the worship of the Syrian goddess Ashima in the capital city.[32] The evidence is not convincing.

The god at Dan in 8:14 is unnamed. He is designated "your god" at Dan. Although this may refer to the personal relationship between the deity and his worshipers, the phrase "your god" in the Book of Amos appears simply to refer to Yahweh (4:12; 9:15). Barstad suggests it designates a local Baal, the city-god of Dan[33]; again the evidence is not convincing.

What can we make of the story concerning the origin of the sanctuary of Dan in Judges 17–18? The verses are clearly a polemic against the legitimacy of Dan as a sanctuary. We are told that the tribe of Dan, unable to secure a place on the Philistine Plain by the Mediterranean Sea, migrated to the northern city of Laish at the foot of Mount Hermon. On the way they seized sacred objects dedicated to Yahweh by an Ephraimite named Micah: an ephod, teraphim, and a graven image made from stolen money. Micah, moreover, had a private priest, a Levite, descendant of Moses' son Gershom, who agreed to leave his patron and become a priest for the tribe of Dan.

It is a curious passage. While it does legitimize the priesthood at Dan by indicating the priests were descendants of Moses (compare I Kings 12:31), the passage also reveals that the sacred site's origin with Micah was false from the beginning, because it possessed a graven image (made from stolen money, hence a double iniquity) as well as an ephod and teraphim, all of which were illegal. Micah's private shrine was also maintained by stolen money, and the Levite who officiated at Micah's shrine was not free, but under Micah's employment.

Micah's image was possibly the bull figure associated with Dan. One would not, of course, expect the redactors to associate the two objects; for, in their judgment, Jeroboam I was the person who led Israel astray by erecting the golden calves at Bethel and Dan. But sources are too legendary for us to be confident of their historical accuracy.

Attempts to interpret "the way to Beer-sheba" in Amos 8:14 are multiple. The Septuagint reads "your god"; this is obviously borrowed from the preceding line. E. Hammershaimb suggests altering the text from *derek* "way" to *dōděkā* "your tutelary deity"[34]; Watts translates *dōděkā* "your beloved"[35]; A. S. Kapelrud translates it "your Dod," a reference to a god named Dod at Beer-sheba.[36] F. J. Neuberg proposes emending to *dōrěkā* "your assembly"; he translates it "your pantheon, Beer-sheba," referring to the assembly of the gods at Beer-sheba.[37] Barstad relates *derek* to the Ugaritic *drkt* "dominion,

strength, might"; he translates it "the might of Beer-sheba."[38]
Whether the verse refers to Yahweh or some foreign deity is uncertain. The translation "the way of Beer-sheba" is, I believe, preferred; the meaning of this translation will be explored later in this chapter.

Did Amos condemn the northern kingdom for worshiping foreign deities? Barstad believes he did, but the evidence is not decisive. A fairly strong argument can be made that 2:8b and 8:14a refer to foreign deities. The argument is equally convincing, however, that Amos was attacking a syncretistic form of Yahweh worship. This does not deny that the northern kingdom was engaged in Baal worship. (The Book of Hosea contains a scathing attack on Baalism as practiced in the north.) But it does raise once again the question of why Amos remained silent on this issue.

Amos' Rejection of Worship at Northern Cultic Shrines

We have seen that it is doubtful Amos believed ethical obedience to God's will was the sole form of worship. He did not advocate a cultless religion. On the other hand, he opposed in the strongest terms the substitution of cultic worship for ethical obedience. This substitution God abhorred. Nor was Amos tolerant of the many Canaanite influences within the northern cult. However, it is not clear from the text that he thought the north worshiped foreign gods. He may well have believed the northern cult practiced an illegitimate form of Yahweh worship.

One thing is clear: Amos opposed worship at northern cultic shrines. His oracles do accentuate a total rejection of northern sites as legitimate places to conduct worship. Five northern cult centers are condemned: Bethel is mentioned by name five times (3:14; 4:4; 5:5, 6; 7:10),[39] Samaria three times (4:1; 6:1; 8:14), Gilgal three times (4:4; 5:5), Beer-sheba twice (5:5; 8:14),[40] and Dan once (8:14).

The three criticisms of Judah are later additions to the text.

1. Amos 2:4–5 is considered a deuteronomistic addition, applying the message of Amos to Judah long after the fall of Israel.[41]
2. Amos 6:1a ("Woe to those who are at ease in Zion") creates special problems. Weiser translates the phrase "those who are proud of Zion," referring to the pride Israel took in the capture of Jerusalem following a victory at Beth-shemesh (II Kings 14:11–14).[42] The text does not support such a translation. More likely

6:1a is also a deuteronomistic addition, warning those in Judah to take care lest they commit the same errors that led to the destruction of Israel.[43] There is no further mention of Judah in chapter 6 of Amos, which supports 6:1a as an intrusion into the text.

3. The same approach applies in 3:1b in which the condemnation of the north is expanded to include the south as well.[44] Or the verse may be a northern cultic election saying that Amos parodies.

Why did Amos not condemn worship at southern shrines but opposed it at northern shrines? We saw in the preceding chapter that Amos criticized the surrounding foreign nations for rebelling against the Davidic empire. Israel also rebelled against Davidic rule. Not only did that kingdom establish a monarchy opposed to the Davidic line, but it also possessed its own capital, its own priesthood, a separate festival calendar, and its own sacred sites. We have also seen that throughout the Ancient Near East, priesthood and monarchy supported one another. This was true in Judah; it was also true in Israel. This is one of the reasons the classical prophets criticized priests and kings alike; the priesthood and the monarchy often collaborated to form a state religion that served only their own selfish ends. This collaboration led to corruption in the cult and injustice in the courts. The classical prophets did not, however, reject the priesthood and the monarchy per se.[45] They called for the priests to instruct the people in the ways of God and for the kings to establish justice in the courts.

Amaziah, priest at Bethel, is a good example of a loyal representative of the crown. He understood that Amos' attack on Bethel was at the same time an attack on the monarchy (7:10–17). When Amos declared that Bethel and Gilgal would be destroyed (5:5), it meant the end of the northern kingdom (7:11). With his attack on the northern shrines, Amos struck at the foundation of the northern monarchy. Little wonder that Amaziah prohibited Amos from further prophesying at Bethel, insisting that he return to his own land. For Amos, the only way true religion could be reestablished in Israel was for that nation to reunite with Judah, rejecting both northern cult and monarchy.

Amos' Polemics against Northern Shrines[46]

The validity of this interpretation of Amos' attack on cultic worship has to be judged by the light it sheds on the text. We turn therefore to an examination of those Amos oracles critical of northern religious practices.

Bethel

The northern shrine that figures most prominently in the Book of Amos is Bethel. This may be ascribed partly to what H. W. Wolff calls the Bethel editor, who was influenced by Josiah's destruction of that site during his religious reforms (II Kings 13:1–14; 23:15–18). This editor reworded Amos' oracles against Bethel and included details of the site's destruction. The editorial changes are not difficult to identify.[47] We shall note three examples.

1. Amos 3:13–15 is an oracle in the form of an "installation of witnesses."[48] Amos' anticultic message is in 3:14b:

> And the horns of the altar shall be cut off
> and fall to the ground.

The coming destruction of the northern kingdom will be total. Even the horns of the altar that offer asylum for a fugitive will be destroyed (Exod. 21:13–14). The guilt of the northern kingdom is so great that a place of refuge will be useless. A form-critical analysis reveals that 3:14ba, "I will punish the altars of Bethel," intrudes into the oracle.

2. According to Wolff, Amos 5:6 provides a second example from the hand of the Bethel editor. Amos 5:4–5 is a genuine oracle of Amos condemning Bethel, Gilgal, and Beer-sheba;[49] v. 6 is the Bethel editor's expansion of the oracle to include Bethel's destruction:

> Seek the LORD and live,
> lest he break out like fire in the
> house of Joseph,
> and it devour with none to quench it
> for Bethel.

Wolff argues that the northern kingdom is called the "house of Joseph" only after its destruction as a nation in 722/1. Later editors refer to Joseph two other times in the Book of Amos, all related to the destruction of the northern kingdom (5:15, "the remnant of Joseph" and 6:6, "the ruin of Joseph").

3. The Book of Amos contains three so-called doxologies or par-

ticipial hymns of praise (4:13; 5:8–9; 9:5–6). While they are similar in structure and content, enough differences exist to justify their being considered three separate hymns. They contain Canaanite cultic language that has been reworked to form a polemic against foreign religious influences. After praising the deity's activity in the world of nature and among humankind, each hymn insists that the object of its praise is Yahweh, not some other god. While other people may worship their gods, the hymns proclaim that Yahweh is the true creator and sustainer of life. Amos 4:13 provides a clear example.

> For lo, he who forms the mountains and
>   creates the wind,
>   and declares to man what is his thought;
> who makes the morning darkness,
>   and treads on the heights of the earth—
> the Lord, the God of hosts, is his name!

The function and placement of these hymns in the Book of Amos are not clear. But an attractive theory holds that they are related to the destruction of the altar at Bethel by Josiah. It is even possible that they were once used at Bethel in praise of some foreign deity, but are now employed in the worship of Yahweh.[50] Wolff argues persuasively that each hymn is placed over against a genuine Amos' oracle condemning Bethel worship: 4:13 against 4:4–5[51]; 5:8–9 against 5:4–5[52]; and 9:5–6 against 9:1–4.[53] He suggests these hymns are part of a liturgy spoken at the destroyed altar at Bethel glorifying Yahweh, a type of judgment doxology.

Bethel is an old Canaanite site, dedicated, as the name indicates, to the god El. Located ten miles north of Jerusalem, it was founded ca. 2000 B.C.E. Aside from Jerusalem, it is the most frequently mentioned city in Hebrew Scripture. Excavations indicate that the site was occupied as early as ca. 3200. According to patriarchal traditions, both Abraham (Gen. 12:8; 13:3–4) and Jacob (Gen 28:19; 35:1–7) were associated with Bethel; this may be accurate, but the patriarchs would then be worshiping the god El. Not until the Late Bronze Age did the name El become one of the several terms for Yahweh. Excavators in 1960 discovered a sanctuary on the acropolis dating primarily from the Middle Bronze Age I period (2100–1900 B.C.E.). It was probably destroyed by earthquake.

During the settlement period, Bethel was captured and apparently occupied by the Hebrew people. Whether the capture of Bethel was occasioned by an external invasion or by an internal peasant's revolt is presently a debated question. That it became a Hebrew city is unquestioned. It probably housed the ark of the covenant for a time during the premonarchical period (Judg. 20:18–28). Bethel achieved new prominence under Jeroboam I when it became the northern kingdom's chief sanctuary.[54] According to the deuteronomists (I Kings 12:28–33) and perhaps Hosea (8:5), it possessed a golden calf, but archaeological excavations have failed to locate the calf sanctuary.

There is some archaeological evidence that the site was destroyed when Samaria fell to the Assyrians. If this were the case, the shrine was revived toward the close of the Assyrian period and became a major worship center heavily influenced by Canaanite practices. According to the deuteronomistic historians, Josiah destroyed the sanctuary in 621, but J. L. Kelso's archaeological report indicates that the city remained occupied until the time of Nabonidus (ca. 540) or until the early Persian period.[55] The deuteronomistic writers may have been referring only to the destruction of the temple. Archaeologists, however, have failed to locate any temple site at Bethel.

With the exception of those passages related to the *marzēaḥ*, the most detailed verses in Amos describing the nature of cultic worship in Israel are 4:4–5 and 5:21–24. Although one cannot determine whether the oracles were proclaimed at Bethel, Gilgal, or Samaria, it seems appropriate to analyze their content at this time. Worship at Bethel probably differed little from that practiced at other shrines. The technical terminology of the various cultic acts is identical for worship in both the north and the south. To the outside observer, what distinguishes the south from the north is only the sites where worship took place.

Amos 4:4–5 reads:

> "Come to Bethel, and rebel;
>   to Gilgal, and multiply rebellion;
> bring your sacrifices in the morning,
>   your tithes on the third day;
> offer a sacrifice of thanksgiving of
>   that which is leavened,
>   and proclaim freewill offerings,
>   publish them;
>   for so you love to do,

O people of Israel!"
says the LORD God.
(author's
translation)

The passage is a parody of a priestly instruction. While the priests called the worshipers to "come" to the sanctuary and kneel before God (Ps. 95:6), rendering thanks and praise (Ps. 100:4), Amos called them to "come" to Bethel and Gilgal where they rebel against God. The passage is filled with irony. Here alone Amos used the verbal form of *peša'*, indicating that the pilgrimages to Bethel and Gilgal are an act of rebellion against Yahweh. Wolff differentiates between the noun, which connotes a "crime against human community," and the verb, which designates an act of rebelling against God. He argues that the verbal form in this passage should be given its customary meaning of infraction of personal and property rights.[56] J. L. Mays rightly perceives that the verbal form signifies "a break with the God whose community they sought."[57] The NEB correctly makes the distinction between "rebel" (4:4) and "crime" (1:3). Whereas the purpose of cultic rites is to overcome the gulf between the worshipers and the deity, the rites performed at these sites widen the gap between the people and Yahweh. Each act of worship offends God; the people may love to perform the rites, but according to Amos God does not desire them.

These verses contain four types of cultic practices:

1. The pilgrim sacrifices an animal to God that culminates in a banquet meal in which the meat is consumed. Just as fellowship with other people emanates from a shared meal, so the worshiper experiences a special relationship with the divine through eating the meat that has been sacrificed. While this may occur "every morning" (RSV), more likely the phrase means "in the morning," that is, in the morning after arrival at the place of pilgrimage.
2. The tithes the worshipers bring are probably the tenth portion of the harvest—grain, wine, and oil—although they might also include the flock. The tithes are most likely offered not "every three days" (RSV), but rather "on the third day" of arrival at the sanctuary. In this way God receives a share as a gift, while the people enjoy the remainder after first honoring the deity who has blessed them.

3. Thanksgiving offerings of that which is leavened are burnt; the reference perhaps is to the rings of bread from leavened dough (Lev. 7:13).
4. Freewill offerings are those gifts brought from an inner yearning to honor God, not from the incentive to redress some offense.

The priestly instruction included explaining that God desires such practices. In biting contrast, Amos proclaimed that the people themselves are the ones who love offerings and sacrifices, because they enjoy enhancing their own reputations by publishing aloud what they offer. Amos declared to the people that "your" sacrifices and "your" tithes are the ones being offered, not God's.

Amos 5:21–24 contains an even more detailed list of cultic rites that the people offer God. In language used only to condemn evildoers and idolaters, Amos declared that God "hates and despises" these rites. They offend all of God's senses: he will not smell them (v. 21b), look upon them (v. 22b), or listen to them (v. 23).

> I hate, I despise your feasts,
>   and I take no delight in your solemn
>   assemblies.
> Even though you offer me your burnt
>   offerings and cereal offerings,
>   I will not accept them,
> and the peace offerings of your fatted
>   beasts
>   I will not look upon.
> Take away from me the noise of your songs;
>   to the melody of your harps I will not listen.
> But let justice roll down like waters,
>   and righteousness like an ever-flowing stream.

God condemns every element of their worship.[58]

1. Festivals (v. 21). The festal solemn assemblies might refer to the three great pilgrimage feasts: the Feast of Unleavened Bread, the Feast of Weeks, and the Feast of Tabernacles. God will not smell sacrifices offered on these occasions. The RSV translates "I take no delight in"; Wolff's translation, "I will not savor," is more appropriate.
2. Sacrifices (v. 22). The burnt offering is totally consumed on the altar. The verse makes a distinction between the burnt offering

and the peace offering (*šelem*) in which a part of the animal is burned and the rest is consumed by the worshiper. Peace offering (from *šālôm*) is probably an inaccurate translation; the name of the offering likely derives from the verb meaning "to complete," hence a banquet held at the completion of the cultic celebration.[59] The cereal offering is a bloodless sacrifice; it is a term used collectively for an offering in general. These are the sacrifices God refuses to notice.

3. Songs of praise (v. 23). The songs and the harps produce a disagreeable noise that is offensive. Amos may have had in mind here an aspect of the *marzēaḥ* (see 6:5). In any case, God refuses to listen to the din of music.

Amos used a technical cultic term when he declared that these sacrifices are not "accepted" by God (v. 22). It is the same word used by the priests when a sacrifice is not properly performed (Lev. 19:5–8). What makes these sacrifices unacceptable? Certainly the people's failure to practice justice and righteousness offends God (v. 24). But God also rejects sacrifices offered at sites unacceptable to him, sanctuaries supported by a defiant people led by a rebellious government (4:4–5; 5:4–5).

Gilgal

Gilgal is mentioned three times in the Book of Amos (4:4; 5:5); both passages are considered genuine Amos oracles by the form critics. In both, Gilgal is coupled with Bethel. The first passage, a parody of priestly torah, invites the worshipers to come to Gilgal and thereby increase their rebellion against God. The second reference, though resembling a parody of priestly torah, is less ironic in tone and might better be classified as a prophetic exhortation.[60] It warns the worshipers not to pilgrimage to Gilgal because its cultic personnel will be exiled.

The name Gilgal means "circle of stones." There are a number of Gilgals mentioned in Hebrew Scripture, but the major cultic site referred to by Amos was undoubtedly the one located two and one-half miles east of Jericho. Excavations in 1954 uncovered remains from the Early Iron and Middle Iron periods (ca. 1200–600). Here archaeology supports the literary traditions that give the site prominence during the age of Samuel and Saul. The Gilgal mentioned in the Elijah–Elisha stories is probably another site located seven miles northwest of Bethel.

Joshua 3–5 contains many religious traditions associated with Gilgal. As in the crossing of the sea, the Jordan River divided when the ark of the covenant preceded the Hebrew people into the promised land. Twelve memorial stones taken from the Jordan were rolled *(gālal,* hence the name Gilgal) into a circle to commemorate the event. Gilgal became the first encampment in Canaan; from there came the conquest of the central region. We are told that at Gilgal the people were circumcised (a story perhaps also occasioned by the Hebrew word *gālal* "to roll away"), a Passover was held, and Joshua received a theophany. At Gilgal, Joshua made a treaty with the Gibeonites, which eventually led to the defeat of the Canaanite southern confederacy. Legend and history have obviously been blended in the Gilgal traditions.

An annual cultic festival may have been celebrated at Gilgal where there were reenactments of the miraculous crossing of the sea and of the entry into the promised land. A recital of the exodus from Egypt and of the entrance into Canaan undoubtedly accompanied the cultic acts.[61] Psalm 114 preserves the community's memory of this festival:

> When Israel went forth from Egypt;
> > the house of Jacob from a people of
> > > strange language,
> Judah became his sanctuary,
> > Israel his dominion.
> The sea looked and fled,
> > the Jordan turned back.
> The mountains skipped like rams,
> > the hills like lambs.
> What ails you, O sea, that you flee?
> > O Jordan, that you turn back?
> O mountains, that you skip like rams?
> > O hills, like lambs?
> Tremble, O earth, at the presence of
> > the LORD,
> > at the presence of the God of Jacob,
> who turns the rock into a pool of water,
> > the flint into a spring of water.

On a more historical plain, Samuel judged cases at Gilgal and there Saul was elected king by the populace (I Sam. 7:16; 11:14–15);

the site continued as a leading cultic center during the monarchy. In addition to the Amos' references, the Book of Hosea denounces Gilgal three times (Hos. 4:15; 9:15; 12:11).

Samaria

King Omri of Israel (876–869) founded Samaria as the capital city of Israel. It was located in the central region forty-two miles north of Jerusalem. Though lacking a good water supply, the hill is prominent enough to be easily defended. The site has been excavated on four separate occasions: in 1908–1910 by Harvard University, in 1931–1935 by a joint expedition of five institutions, in 1965 by the Department of Antiquities of Jordan, and in 1968 by the British School of Archaeology in Jerusalem.

The remains on the hill's summit of the Omri–Ahab city indicate a well-designed and beautifully decorated structure. Wall fragments from two main systems of fortifications evince the skill of Phoenician masons. Inside the walls were found the royal sections of the city, which consisted of a number of buildings and courtyards and included a temple with adjacent courtyard. This temple was presumably built by Ahab and dedicated to Baal (I Kings 16:32). A shallow pool was in the northwest corner of the complex. Apparently most of the construction of the royal city occurred under Omri and Ahab; Jehu and his immediate successors may have repaired the wall, rebuilt earlier structures, and erected a few new buildings. The next major period for repair and reconstruction was under Jeroboam II.

The two most interesting finds are the ivories and the Samaria ostraca. Excavators attribute the ivories to the Omri–Ahab period; these consist of more than 500 fragments from wooden wall panels and furniture inlays. Their style suggests they were carved by Phoenician artisans. Among the figures are winged cherubim, lions fighting with bulls, and such Egyptian gods as Horis, Isis, and Ra. These ivories indicate a capital city of considerable wealth that welcomed outside cultural influences. These finds add meaning to the biblical references concerning ivory houses and beds at Samaria (I Kings 22:39; Amos 3:15; 6:4).

The Samaria ostraca consist of sixty-three potsherds inscribed in Hebrew recording shipments of oil and wine by various settlements in the area to the royal house for payment of taxes. They have been dated from as early as the reign of Ahab (869–850) to as late as the reign of Menahem (745–737), but the reigns of Jehoash (802–786) and of Jeroboam II (786–746) are most favored.[62] Their greatest contribution is in the areas of Hebrew paleography and the study of personal names and state administration. The ostraca indicate an active and prosperous economy.[63]

The Book of Amos contains a number of passages that appear to denounce cultic worship in Samaria. We have already noted that 2:7b–8; 4:1–3; 6:4–7 may refer to a *marzēaḥ* festival celebrated in Samaria. The temple mentioned in 8:3 may refer to the sanctuary in Samaria (see 8:14), but the reference lacks definition. Possibly Amos was condemning idolatry in 6:13, but that is not the most favored interpretation of the text.[64] Most critics believe 9:1–4 refers to Bethel and not Samaria. But as in the case with Bethel and Gilgal, Amos declared that Samaria will be destroyed and the people exiled (3:9–15; 4:2–3; 6:7–14; 8:9–14).

Sargon II captured Samaria in 722/1 and burned part of the city. It was reconstructed by the Assyrians and repopulated with political exiles from other regions of the Fertile Crescent. Under the Assyrians, the Babylonians, and the Persians, Samaria served as an administrative center. Herod the Great rebuilt and renamed it Sebaste in honor of Emperor Augustus. It existed under that name until Arab occupation in the seventh century c.e.

Dan

Dan sits at the southern foot of Mount Hermon on the trade route between Tyre and Damascus. The city was originally called Laish before it fell to the tribe of Dan in the settlement period. Excavations have been conducted at the site by the Department of Antiquities and Museum of Israel every year since 1966.

The earliest period of occupation thus far detected is Early Bronze Age IV (2800–2600 B.C.E.). Laish was then a large and prosperous city; excellent springs nearby supplied abundant water and form one of the tributaries of the Jordan River. During the Middle Bronze Age II B period (1700–1600), the site was strongly fortified with massive ramparts. Artifacts from this period indicate a thriving city that was mentioned in Egyptian Execration Texts and Mari inscriptions. A Mycenaean tomb from the Late Bronze Age II period (1300–1200) contained imported Mycenaean and Cypriot wares as well as fine domestic articles. A destruction stratum dated the middle of the eleventh century probably marks the tribe of Dan's occupation of the site. There is every indication the community prospered throughout the period of both the united and divided kingdoms. Conquered by the Assyrians in the eighth century, the city continued to be occupied as late as the Byzantine period.

The two most distinctive features from the period of Israelite occupation are the city wall with its gate and a high place for open-air worship. The massive twelve-foot-thick city walls were probably built by Jeroboam I when he designated Dan as one of the two leading cities of his kingdom. The large worship platform approached by a wide flight of steps was probably also erected by Jeroboam I and expanded by both Ahab and Jeroboam II. According to biblical accounts, Jeroboam I placed a golden calf at Dan, but no evidence for its existence has been uncovered. However, an Israelite horned altar, similar to the one found at Beer-sheba, was discovered during the 1974 season.[65]

The reference to Dan in the Book of Amos (8:4) offers no insights into the nature of worship at this far northern site. It is doubtful that Amos journeyed beyond Samaria to prophesy at Dan. The phrase "your god(s)" may refer to Baal worship, but could just as easily refer to an unacceptable form of Yahweh worship. Amos did condemn worship at the site of Dan, proclaiming its imminent destruction.

Beer-sheba

Located twenty-eight miles southwest of Hebron, Beer-sheba was the principal city in the Negeb. Beer-sheba figured prominently within the patriarchal narratives. Two references that explain the name differ in detail. In Genesis 21:25–31, Abraham there gave to Abimelech of Gerar seven lambs and named the place Beer-sheba, "well of the seven." But according to Genesis 26:32–33, when Isaac made an oath there to seal an alliance with Abimelech, he called the place Beer-sheba, "well of the oath." In his old age, Jacob visited Beer-sheba on his journey to Egypt (Gen. 46:1–5). According to Genesis 21:33 the Canaanite god El Olam was worshiped there. If this is accurate, the patriarchs would have worshiped El Olam at Beer-sheba; later this deity was assimilated into Yahweh worship and his name became a divine epithet meaning "the Everlasting God."

Recent excavations have disclosed a well-planned Israelite city established during the united monarchy. The city possessed two major wall systems. The massive solid wall with a moat at the base of a glacis is early; that wall was replaced by a casemate wall, which was destroyed in the eighth century. The city possessed well-planned streets, typical four-room homes, and an impressive gate system. There were long rooms used for storehouses, rooms similar to those discovered at Megiddo.

Two finds have religious significance. First, excavators discovered the remains of a cultic platform with an incense altar nearby. The platform is remarkably similar to the open-air altar at Dan. Second, in excavating the storehouses, workers discovered that a section of the walls had been restored from pieces of a destroyed horned altar, apparently one of the altars dismantled by Hezekiah during his religious reforms (II Kings 18:22).[66]

The city was destroyed about the end of the eighth century, probably by Sennacherib's campaign in 701.[67] The settlement then went unoccupied until the Persian period when it was probably made into a fortress. The city continued to have military significance during the Hellenistic and Roman periods.[68]

The two references to Beer-sheba in the Book of Amos are remarkably similar. In 5:4 the northern worshipers are instructed not to "cross over to Beer-sheba," and in 8:14 those who swear by "the way of Beer-sheba" are condemned. Both references indicate that Beer-sheba was a pilgrim site sacred to the northern kingdom. The biblical evidence is extensive that Beer-sheba was a northern shrine. The area figures prominently within the northern tradition about Isaac and Jacob. According to I Sam. 8:2, Samuel's sons from the northern tribe of Ephraim served as judges in Beer-sheba. I Kings 19:3–4 records that the northern prophet Elijah visited Beer-sheba on his flight to Horeb.

Earlier in this chapter, we examined a number of suggested emendations for the reference to *derek* ("the way") in 8:14.[69] We would do well to leave this text unchanged. The conclusion is sufficiently clear: just as Amos condemned the northern shrines of Bethel, Gilgal, Samaria, and Dan, so also he condemned pilgrimages to that other site sacred to Israel, Beer-sheba. Wolff argues that in 5:4–5 the phrase "cross over to Beer-sheba" intrudes into the text. The worshipers are instructed not to seek Bethel and Gilgal, for these two shrines will be destroyed. Because there is no mention of Beer-sheba's destruction, Wolff finds the unity of the passage disrupted.[70] But the sequence forms a perfect chiasma: Bethel–Gilgal–Beer-sheba–Gilgal–Bethel. Furthermore, the reason Beer-sheba is not marked for destruction is that it lies within the borders of Judah. Amos never proclaimed the destruction of Judah. Amos 8:14 confirms this interpretation: It is *not* the sanctuary of Beer-sheba that will fall, but rather those northern persons who pilgrimage there.

Amos and Jerusalem

Until recent years, biblical scholars have maintained that the concept of worship centralized in Jerusalem originated with the Josiah reforms of 621.[71] The Scythian invasion of 626 had demonstrated the weakness of Assyrian power. In a few years Assyria would fall before the combined forces of the Babylonians, the Medes, and the Scythians. Josiah seized the opportunity to free Judah from Assyrian dominance; one of his first acts was to purge pagan practices from the Jerusalem temple. During the cleansing process, an earlier law code was found that revealed how far from pure Yahwism the community had strayed. Josiah made this law code the basis of his reforms. He extended his influence into the northern regions and he dismantled various high places, the most prominent of which was Bethel. He made Jerusalem the only legitimate place of worship in Canaan. The Levitical priests associated with the provincial sanctuaries became lesser officials in the Jerusalem cult. The long-neglected Passover was made the major festival.

The religious reforms of Josiah should not be minimized. Although the deuteronomistic historians were prejudiced in favor of Josiah (regarding him and Hezekiah as the best kings religiously who ever ruled Judah), Josiah still deserved their praise. His was the most effective attack on pagan practices ever instituted in Judah.

To argue, however, that this was the first successful cultic reform in Judah is to disregard the contributions of several leading southern kings. Shortly after the split of the kingdom, Asa (913–873) attempted to eliminate the pagan influences introduced by Rehoboam.

> In the twentieth year of Jeroboam king of Israel Asa began to reign over Judah, and he reigned forty-one years in Jerusalem. His mother's name was Maacah the daughter of Abishalom. And Asa did what was right in the eyes of the LORD, as David his father had done. He put away the male cult prostitutes out of the land, and removed all the idols that his fathers had made. He also removed Maacah his mother from being queen mother because she had an abominable image made

of Asherah; and Asa cut down her image and burned it at the brook
Kidron. But the high places were not taken away. Nevertheless the
heart of Asa was wholly true to the Lord all his days.

(I Kings 15:9–14)

The chronicler records even more extreme reforms; heeding the
prophet Azariah, Asa repaired the altar and persuaded northern
sojourners in the south to worship Yahweh in Jerusalem (II Chron.
14:1–5; 15:8–15). Both the deuteronomists and the chronicler claim
Asa removed the queen mother from her position because she possessed an image of Asherah (I Kings 15:13; II Chron. 15:16).[72] And
Asa's son Jehoshaphat (873–849) continued his father's reforms (I
Kings 22:43; II Chron. 17:3–6). Addressing the historical accuracy of
these passages is difficult. The references in I Kings follow the stereotyped style of the deuteronomists, but the content is probably based
on some factual knowledge of Asa's religious reforms.

A further example of cultic reform occurred when Athaliah,
daughter of Jezebel and queen ruler of Judah (842–837), attempted
to make Baal worship an established religion in Judah (II Kings 11:18;
II Chron. 23:17). Loyal Yahweh priests revolted, had Athaliah put
to death, and installed Joash (837–800) as the legitimate heir to the
throne. Joash then initiated a collection of funds to make extensive
repairs to the temple (II Kings 12:4–16; II Chron. 24:4–14). The
chronicler adds that such repairs were necessary because the sons of
Athaliah had plundered the temple and had used the sacred utensils
for the worship of Baal (II Chron. 24:7). The deuteronomists lament
that although the temple was repaired, Joash erred in not destroying
the high places (II Kings 12:3).

With the exception of Josiah, Hezekiah (715–687) was the greatest
religious reformer among the Judahite kings.[73] The deuteronomists
devoted three chapters to his reign (II Kings 18–20) and the chronicler
four chapters (II Chron. 29–32). Concerning his religious reforms, II
Kings records his removal of high places, breaking of pillars, cutting
down the Asherah, and the destruction of the bronze serpent, which
tradition assigned to the Mosaic period.[74] The chronicler, as was his
custom, added his own embellishments. He devoted three full chapters to the cleansing of the temple, to the place of the Levites, and
to the celebration of the Passover in Jerusalem by both northerners
and southerners. Although some critics argue that the stereotyped
description of the reforms found in II Kings 18:1–6 reflects the period

of Josiah, most critics regard the deuteronomistic references as basically accurate. John Gray writes: "there is no reason to suppose that the nationalistic policy of Hezekiah did not have as its religious aspect the centralization of the cult."[75] The discovery of the dismantled horned altar at Beer-sheba dated from the eighth century appears to confirm the deuteronomistic account. An attractive hypothesis relates the reforms of Hezekiah to those of Josiah by suggesting that the deuteronomic law code reflects the influence of Hezekiah's reign; the code was written during the Manasseh period, then resurfaced under Josiah.[76]

The reason for this recital of religious reforms in Judah from the period of the division of the kingdom down to its final days is to dispel the idea that Josiah was the only advocate in Judah for centralization of worship in Jerusalem. Ever since David had united the ark and the tent in Jerusalem, Zion was the chief religious center in Canaan. During Absalom's revolt, for example, David evacuated Jerusalem but sent the ark back to the capital, indicating that the sacred object should reside nowhere else (II Sam. 15:24–29). Neither Bethel nor Dan had cult objects to rival Jerusalem's popularity in Judah for Judeans.

Amos was a Judahite prophet. What was his attitude toward the Jerusalem cult? Most critics argue that he was equally critical of cultic practices in the south and in the north. The only reason he did not mention sanctuaries in the south was that he was called to prophesy to the north. That hardly addresses the issue. As we have reviewed, his attack on cultic worship was directed *exclusively* against northern shrines. Not one authentic Amos oracle condemns southern religious centers. Why did Amos feel compelled to prophesy exclusively in the north?

The answer lies in two references in the Book of Amos. The first reference is 1:2, which I believe is an authentic Amos oracle:

> The LORD roars from Zion
> and utters his voice from Jerusalem;
> the pastures of the shepherds mourn,
> and the top of Carmel withers.

The verse occurs in modified forms in Jeremiah 25:30 and Joel 3:16. The Jeremiah passage contains a condemnation of all wicked foreign nations. This is similar to Amos' oracles against foreign nations. And while the Joel reference is also a judgment against foreign nations, it is unlike Amos because it contains a hope passage for Israel. Most critics regard Amos 1:2 as a hymnic passage originating within the Jerusalem culture.[77] It bears a likeness to Psalm 50. But there is no reason why Amos could not have reworked the psalm as an appropriate summary of his message. Psalm 50 contains words and imagery strikingly similar to Amos (see Ps. 50:3, 6, 7, 9, 17, 22); the psalm would have appealed to Amos.[78] Furthermore, Amos 1:2 is in the style of other passages accepted as authentic Amos. In 1:2 it is the Lord's voice that is destructive; in 3:8 God's word given to Amos for proclamation is destructive. Yahweh in 1:2 is the divine warrior who comes not to save but to punish (cf. 2:9, 13–16). His powerful voice causes the pasturelands to wilt and the plentiful forests on Mount Carmel within Israel to wither (cf. 7:4–6; 8:13–14; possibly 4:6–8 and 5:16–17). Amos 1:2 is an apt summary of Amos' essential message announcing the total destruction of the northern kingdom. Actually the emphasis on Zion and Jerusalem in 1:2a is even more pronounced than the RSV reveals. A literal translation reads:

> The LORD from Zion roars,
> and from Jerusalem he utters his voice.
> (author's translation)

The word order is significant; it is *from Zion* that the Lord roars and *from Jerusalem* that he utters his voice. By placing the verbs later in the sentence, the emphasis falls on Zion and Jerusalem. The second half of the verse resumes the normal Hebrew word order, placing the verbs first.[79] The verse forms a perfect introduction to Amos' oracles against foreign nations and to his polemic against northern shrines.

The second reference is 5:4–5:

> For thus says the LORD to the house
> of Israel;
> "Seek me and live;
> but do not seek Bethel,
> and do not enter into Gilgal
> or cross over to Beer-sheba;
> for Gilgal shall surely go into exile,
> and Bethel shall come to nought."

The form of the oracle is a priestly torah, instructing the worshipers not to pilgrimage to Bethel, Gilgal, or Beer-sheba, but to seek God and live. Amos purposely chose two technical cultic terms, *dāraš* ("to seek") and *bōʾ* ("to enter"), to warn the worshipers against certain sanctuaries.[80] Concerning the use of *dāraš*, Psalm 24 calls the worshipers to ascend the hill of the Lord and stand in his holy place, for "such is the generation of those who seek him, who seek they face, O Jacob" (v. 6).[81] In Psalm 53 God looks down from heaven to see if there are any persons "that seek after God" (v. 2 [H v. 3]). That God is to be sought in the temple is clear when the psalmist closes with the petition, "O that deliverance for Israel would come from Zion!" (v. 6 [H v. 7]). Similarly, *bōʿ* is used in the familiar call to worship of Psalm 95:6:

> O come, let us worship and bow down
> let us kneel before the LORD, our Maker!

What did Amos intend when he instructed the northern worshipers to seek God and live and not to seek Bethel, Gilgal, or Beer-sheba? Mays argues: "Amos usurps the function of the priests of Bethel by giving *tôrâ* himself in which he replaces shrine with the divine person, and then contradicts the priestly office by forbidding the Israelites to come to the shrine at all."[82] Is it not truer to the text to interpret Amos' words as a call to seek God where he may be found, that is, in Jerusalem?[83] We shall have occasion to explore more fully both the meaning of *dāraš* and the form and intention of 5:4–5 as well as related verses in Chapter 7.

6

Social Justice and the Just King

FOR THE general reader of Hebrew Scriptures, the call for social justice is the essential message of the Book of Amos. Amos is the first classical prophet to champion the cause of the poor and the needy in the land. The verse most often associated with him is 5:24:

> But let justice roll down like waters
> and righteousness like an ever-flowing stream.

Amos condemned corruption in the courts and the marketplaces and called Israel "to establish justice in the gate" (5:15). While some may debate Amos' views of Yahweh's relationship to foreign nations or of God's attitude toward cultic life, there is no doubt that the prophet condemned the oppression of the poor.

Amos' ideal of a just society contrasted sharply with what he observed in Israel. The precise contours of that contrast, however, are indefinite. Nevertheless, recent scholarship exploring social, economic, and political developments in both Israel and Judah has increased greatly our understanding of the society in which he lived. Today the ethical message of Amos cannot be understood properly apart from the socio-political situation of his time.

The Judicial System in the Premonarchical Period

In determining the justice of our own society, we naturally address the efficiency with which the court system administers the law. When we take the same approach with Israel and Judah, our lack of information complicates the task. It is strange that the most sacred texts of a culture—texts that include the five books of Moses with hundreds of laws—seldom divulges how these laws were administered. Only by reading "between the lines," by teasing implicit facts from the text, can one discover information about the judicial system.[1]

The basic social unit of early Israelite society was the "extended family" (*bêt-'āb*), consisting of three or four generations of kin with a common ancestor.[2] The head of the family unit exercised complete judicial authority over the group. Two Genesis texts provide some insight. While these two stories are used by their writers to support specific religious messages for another age, nevertheless they do contain valuable information concerning the judicial system in the premonarchical period. The story of Jacob and Laban in Genesis 31 indicates that once Jacob joined Laban's family unit, Laban had the right to do with him whatever he pleased (v. 29). Furthermore, because Rachel and Leah were Laban's daughters, Laban held that they, their children, and their flocks were still within his jurisdiction; their marriage to Jacob had not altered the situation (v. 43). Jacob, on the other hand, was attempting to form his own family unit over which he had complete authority. The second passage is Genesis 38, the story of Judah and Tamar; the account here reveals that the head of the family had authority to determine whether a family member lived or died. Even after Tamar returned to her own home (v. 11), she was under the authority of her father-in-law Judah. When Judah discovered that the now-widowed Tamar was pregnant, he sentenced her to death (v. 24). When he subsequently learned that he himself was the father of the child, he spared her (v. 26). If, as some scholars maintain,[3] marriage within the Israelite family unit was prohibited, then Judah was guilty of illicit sexual relations with his daughter-in-law. That Judah went unpunished means the head of the family was above the law.[4]

Most scholars maintain the extended family was part of a larger social structure called a "clan" (*mišpāḥâ*). The clan consisted of a number of extended families tracing their lineage to a common ancestor. But because of the significance of the village in Israelite culture, some scholars argue the village formed the next larger social unit after the extended families.[5] The clan–village was administered by the "elders," who were probably the heads of various families. The council of elders, meeting in the cool of the town gate, formed the judicial authority over the clan–village.[6] There disputes among family units would be decided. Communal consensus at these trials was crucial to prevent division. As the town grew in importance, judicial power gradually shifted from the council of the clan–village to the city authorities, who formed the most important judicial court in the premonarchical period. The city council probably continued to exercise judicial authority over cases of the clan–village into the monarchical period. Its exact relationship to the royal justice system is not clear.

A few examples of court proceedings are given in Hebrew Scripture. When Abraham purchased the cave of Machpelah from Ephron the Hittite for the burial of his dead, the transaction took place at the gate of Ephron's city "in the hearing of the Hittites" (Gen. 23:1–16). Or again, when Ruth's closest kinsman was unable to accept the terms of the Levirate marriage vow (see Deut. 25:5–10), Boaz, sitting at the gate of the town in the presence of ten elders, accepted the right of redemption. Ruth was given in marriage to Boaz, and the next of kin underwent what was apparently a humiliating penalty. The ten elders of the town witnessed these transactions (Ruth 4:1–12).[7] In addition to these examples, there are numerous references to elders judging cases (Deut. 19:12; 21:3, 6), often within the city gate (21:19; 22:15). Following such trials, the elders saw that those convicted were punished (21:18–21; 22:18–21).

A number of clans or villages banded to form a tribe. What united the tribes was probably not kinship but rather geographic proximity and common religious and cultural life. The most detailed account of tribal, legal justice in the premonarchical period is the story of the rape and murder of the Levite's concubine in Judges 19–21. A Levite, traveling through the territory of Benjamin, spent the night at Gibeah. Only an Ephraimite sojourner befriended him. During the night certain townsmen demanded to have homosexual relations with the Levite. When this was denied them, they raped the concubine until she died. The chieftains from the various tribes in Israel assembled at Mizpah to try the case. After the Levite testified, the group voted to put to death those who committed the atrocity. When the tribe of Benjamin refused to surrender the guilty parties, the other tribes enforced the court's verdict by military action. They also refused to allow their daughters to marry any Benjaminite, thereby assuring the destruction of that tribe. Following the defeat of Benjamin, however, the tribes regretted their decision to ostracize the rebellious tribe and permitted the men of Benjamin to seize for wives virgin women from two cities that had not participated in the war.[8]

But there are few examples of intertribal attempts to develop a system of justice during the premonarchical period. It is also doubtful that any one judge administered justice over all the tribes.[9] A study of the role of the so-called major and minor judges in the Book of Judges reveals that while they exercised authority over local areas, they had no judicial function or position within any pan-Israelite league. Deborah (Judg. 4:4–5) was a prophetess who exercised authority over the tribe of Ephraim by offering oracular decisions while seated beneath what was called "the palm of Deborah."[10] This hardly qualifies her as a judicial authority. And the annual journey of Samuel from Ramah to Bethel, Gilgal, and Mizpah (I Sam. 7:15–17) was more likely a visit to particular shrines to perform sacral activities rather than an account of Israel's first circuit judge.[11] Furthermore, an examination of the verb *šāpaṭ* suggests that in many cases it bears the more general meaning "to rule," "to govern," "to exercise authority," "to deliver," rather than the specific meaning "to judge."[12]

It does appear, however, that priests were involved in the administration of justice in the premonarchical period.[13] Exodus 22:7–13 [H 22:6–12] contains a series of cases concerning disputed property that were decided by priests at local sanctuaries. The priests, using the Urim and Thummim, determined guilt or innocence.[14] In another case, because there were no witnesses, a woman accused of adultery had to undergo a trial by ordeal administered by a priest (Num. 5:11–31). The decline of the priests' preeminence in judicial matters once the monarchy was established signifies that such practices were part of the premonarchical justice system.

Trial by the casting of lots might also take place in the presence of the clan meeting. This is the method in the story of Achan, Joshua 7. Achan had violated the "ban" by keeping for his private use goods from Jericho that were devoted to God. Through the use of the Urim and Thummim,[15] the thief was discovered. In order to assure the people that justice was done, the record shows that Achan confessed the crime (v. 20) and that his confession was confirmed by disclosure of the stolen goods within his tent (v. 22). The tribes proceeded to execute Achan and destroy all his property, thereby dissociating themselves from what was unholy.

During the premonarchical period, the everyday administration of justice in the local village was apparently under the jurisdiction of clan elders or tribal chiefs who held court "in the gate." When difficult cases arose, they were referred to the local sanctuaries where priests, by the use of the Urim and Thummim or some form of ordeal, discovered God's will.

The Judicial System during the Early Monarchy

The extant law codes of Ancient Israel are attributed to God. Two passages that define the ways of the king (I Sam. 8:11–18 and Deut. 17:14–20) do not attribute to the king any legislative powers. Following a successful military campaign, David did decree that those who guarded the baggage and those who fought in battle should share alike the booty; David's decision did become a "statute" and an "ordinance" in Israel (I Sam. 30:24–25). As R. de Vaux points out, however, David was not then a king; it was as a military commander, with extraordinary battlefield authority, that he acted.[16] Another incident occurred during a siege of Jerusalem when Zedekiah ordered that all slaves be freed. But before making the decision, Zedekiah consulted the people (Jer. 34:8). We have no evidence that the king enacted laws. Gradually, however, the king did develop judicial powers. Against this background of the royal administration of justice, Amos' call for social justice must be set.

As we move into the monarchical period, we would expect the royal judicial system to supplant the older, clan-tribal judicial system. This did not happen immediately, partly because Saul exercised only limited power. Saul was hardly king over more than the tribe of Benjamin. Because of this limited power, the clan-tribal judicial system continued to operate. There is no indication of a developed royal judicial system during Saul's reign. Two major trials were held; both reveal that Saul's authority at the trials stemmed from his position as commander of the army. Both court scenes are military, not civil. Although the passages describing these trials contain many textual, literary, and exegetical problems,[17] our focus will be on the judicial procedures of the day.

The first trial is of Jonathan (I Sam. 14:36–46) following his victory at Michmash (14:1–35). The decision that Jonathan must die was clearly rejected. A close reading of the text suggests that Saul himself was probably guilty of manipulating the trial to serve his own purposes. There are two major points of conflict between Saul and Jonathan.

1. The battle began when Jonathan, accompanied only by his armor bearer, successfully attacked a Philistine garrison while Saul remained in the camp, ignorant of his son's victory. Not only did Jonathan surpass his father as a warrior, but he also made military decisions without consulting the commander of the army.

2. When Saul's army saw that the Philistines were being routed, it joined the battle. During the pursuit, Saul, hoping to secure God's blessing, declared a fast. This resulted in double tragedy. The army, physically weakened by fasting, had limited success in the battle; Jonathan, uninformed of the fast, unwittingly broke it. Jonathan's reaction to his father's declaration of a fast during battle reveals the hostility between them.

> My father has troubled the land; see how my eyes have become bright, because I tasted a little of this honey. How much better if the people had eaten freely today of the spoil of their enemies which they found; for now the slaughter among the Philistines has not been great.
>
> (14:30–31)

Jonathan accused his father of "troubling" the land: he used a word that "appears to imply recklessness or criminal behaviour which exposes individuals or the whole society to the danger of civil and divine revenge."[18]

When Saul and his priest failed to get a favorable sign from God to continue the attack on the Philistines, a trial ensued. The priest revealed through the Urim and Thummim[19] that Jonathan's sin in breaking the fast had offended God. Jonathan was sentenced to death. However, the clan and tribal heads intervened and rescued him.[20] The older clan-tribal legal system was still operative.[21]

A second example of a judicial court scene is at the trial of Ahimelech and the priests of Nob (I Sam. 22:9–19).[22] The passage containing the scene gives a full description of the judicial process. The priests of Nob, apparently appointees of the crown, were summoned to appear before Saul and charged with aiding David in his rebellion against Saul.[23] Doeg the Edomite, chief of Saul's herdsmen (21:7 [H 21:8]), was an eyewitness to the event. He accused Ahimelech and his priestly family of providing David with provisions and a sword and of consulting an oracle on his behalf (22:9–10).[24] The head priest of Nob, Ahimelech, readily admitted to the charges but argued that he was unaware of David's treachery (v. 14). Saul refused to believe him and promptly sentenced to death Ahimelech and all the other priests of Nob. When the king's servants—probably the chief leaders of the people—refused to execute what they regarded as an unjust court ruling, Doeg the Edomite obeyed Saul and destroyed the city of Nob along with the priests, their families, and their property (vv. 18–19). The magnitude of the destruction suggests Saul treated this as a major rebellion in support of David.

In both these court cases the accused were charged with seriously challenging the authority of Saul. In both cases Saul was engaged in warfare, which gave him special authority over a military court.[25] In both cases there was resistance to carrying out the sentence. These cases may simply reflect the pro-Davidic nature of the narrative. But they may demonstrate further not only that kings were capable of abusing their power, but also that the local court system was still viable.

In accounts from the reign of David, the parable of Nathan (II Sam. 12:1–15) and the case of the widow of Tekoa (14:1–24) may appear to reveal something about royal legal procedures. But as K. W. Whitelam demonstrates, both cases are actually literary constructions that serve the purposes of the writers; it is doubtful that either case ever occurred and doubtful that either Nathan or the Tekoite woman ever argued them before the king.[26] Both cases raise too many problems regarding historical accuracy.

The following considerations support the fictitious construct of Nathan's parable.

1. II Sam. 12:1–15a is quite likely a literary insert into the text; 11:27 and 12:15b form a unit. Also 12:1–7a and 13–15a form a unit and vv. 7b–12 are an insert. Nathan's parable has little to do with the sin of David and Bathsheba.
2. Nathan declared that the men came from "a certain city" (v. 1). There is nothing to indicate the case was within the king's jurisdiction.[27]
3. The relationship between the king's judicial authority and the local judicial authority is never clarified.
4. No evidence suggests that Nathan held an office of judicial mediator that authorized him to bring cases before the king.[28]
5. There is no legal warrant for the king to declare that the man deserves the death sentence for what he had done (v. 5).

It is probable that the narrative is a literary construction. Its purpose may be to show that, although the king was above the law and could not be brought to trial for the murder of Uriah, he was still subject to divine justice.[29] David's confession and repentance reveal he was basically a good king; the death of the child demonstrates that God has punished the adulterous affair with Bathsheba, thus removing any curse on the dynasty.[30]

Nor does the case of the widow from Tekoa shed much light on royal legal procedures. Several factors in that case also suggest a literary construction.

1. There is almost no correspondence between the widow's case and David's exile of Absalom. To be sure, Absalom had slain his brother Amnon; and the widow's son had killed his brother. Here the similarity ends. The widow argued for the perpetuation of her family; the return of Absalom from exile has nothing to do with the continuation of the Davidic line.
2. The motives for the two crimes are totally different. Absalom committed a premeditated murder to avenge his sister Tamar. The widow's son acted from sheer anger.
3. It is legally unlikely that a new ruling by the king, setting aside the decision of the clan council (v. 7), would be made retroactive to apply to the Absalom case.
4. In no other instance in Hebrew Scripture does a court defer judgment because the case was too difficult to decide (v. 8).
5. There are two references to the king: the king is the angel of God who discerns good and evil (v. 7); the king has wisdom to know all things on earth (v. 20). These two references reflect the influence of the Solomonic conception of monarchical judicial authority (identified later in this chapter). It is not probable that this view of judicial authority existed during David's reign.

The purpose of this story of the widow from Tekoa is to explain the return of Absalom to David's court. The story provides almost no valid information about the royal judicial court system of the day. That the story is told at all may indicate that at least in theory the royal judicial court had the power to overrule a decision of a local court.

The incident that sheds the most light on the royal judicial system during the early monarchy is the rebellion of David's son Absalom.

After this Absalom got himself a chariot and horses, and fifty men to run before him. And Absalom used to rise early and stand beside the way of the gate; and when any man had a suit to come before the king for judgment, Absalom would call to him, and say, "From what city are you?" And when he said, "Your servant is of such and such a tribe in Israel," Absalom would say to him, "See, your claims are good and right; but there is no man deputed by the king to hear you." Absalom said moreover, "Oh that I were judge in the land! Then every man with a suit or case might come to me, and I would give him justice." And whenever a man came near to do obeisance to him, he would put out his hand, and take hold of him, and kiss him. Thus Absalom did to all of Israel who came to the king for judgment; so Absalom stole the hearts of the men of Israel.   (II Sam 15:1–6)

This was a major rebellion against David in which the entire nation participated. Absalom was crowned at Hebron where his father had once ruled as king of Judah (15:7–10), and David was forced to flee to the Transjordan area because he lacked support in the north.[31] Shimei's curse reflects the extent of hostility in Saul's family toward David:

And Shimei said as he cursed, "Begone, begone, you man of blood, you worthless fellow! The LORD has avenged upon you all the blood of the house of Saul, in whose place you have reigned; and the LORD has given the kingdom into the hand of your son Absalom. See, your ruin is on you; for you are a man of blood."   (16:7–8)

Absalom's conduct at Jerusalem's gate announced his kingly aspirations: He appeared as a warrior with chariot, horse, and fifty men; he also promised to render quick justice in the courts. Both Absalom's show of physical authority and promise of justice were too major characteristics of monarchy in Egypt, Mesopotamia, and Canaan (see Chapter 2), as well as in Judah and Israel (see Chapter 3).

Two points concerning the courts of this time need emphasis. First, although the type of case coming to the royal courts is not clear, those courts probably did not hear appeals from the local city courts. The cases of "appeal" (i.e., Nathan's parable and the story of the widow of Tekoa's son) cannot be trusted for accurate information about the court system.[32] Most likely the cases brought to Jerusalem concerned taxation, management of the king's estates, and other such royal matters, cases that were not under the jurisdiction of village courts.[33] Second, Absalom promised to appoint persons to hear cases brought to the royal court (v. 3). This would constitute a major judicial reform. There is no indication the reform occurred until the reign of Jehoshaphat a century later. Apparently Absalom's defeat discontinued all the promised reforms of the royal court system, because there is little indication that David ran a just court system.[34]

The situation does not improve under Solomon. The narrative presents Solomon as the legitimate heir to David's throne, but it is not successful in hiding that this throne was secured through the manipulation of the court system.

Circumstances surrounding the execution of Adonijah (I Kings 2:13–25) offer fairly conclusive evidence that this was a contrived judicial murder. Adonijah was David's eldest son, the heir apparent (2:15, 22).[35] Nathan and Bathsheba supported Solomon and accused Adonijah of attempting to seize the throne before David's death. The charge is particularly suspect for its blatant use to secure the throne for Solomon. After Solomon became king,[36] Adonijah asked Bathsheba to intercede for him in securing David's favorite concubine Abishag. This story also lacks credulity, because Adonijah could hardly expect help from Bathsheba. Furthermore, the request for a deceased king's concubine is tantamount to asking for the throne. If Adonijah was guilty of this, he should have been executed—for stupidity! These maneuvers were more likely court intrigue to eliminate Solomon's chief rival. Adonijah is sentenced to death on false testimony of one witness, Bathsheba, a witness who was hardly impartial. The whole incident suggests corruption in the court.

Those who had supported Adonijah were also removed from the scene. Apparently Solomon acted within the law to exile the priest, an appointee of the crown (2:26–27). And Joab was legally executed for having unjustifiably murdered Abner and Amasa (2:28–35), two rival commanders of David's army.

Although Shimei avoided supporting Adonijah (1:8), his pro-Saul position (II Sam. 16:5–14; 19:18–23) eventually led to his death (I Kings 2:36–46). In the case of Shimei, the reader suspects wrongdoing. He was confined to Jerusalem and forbidden to leave the city on penalty of death, probably so that Solomon could keep him under surveillance. When he forgot (?) and went to Gath in search of two runaway slaves, Solomon promptly had him executed.

Embedded in this material about the early monarchy's judicial heavy-handedness is a unit presenting Solomon as the ideal king

(I Kings 3:4–28). The passage was probably composed within the Solomonic court for the purpose of legitimizing Solomon's rule. In the form of a dream, the revelation shows Solomon as a humble youth, reluctant to accept the burden of leadership. He prays to become a good king.

> Give thy servant therefore an understanding mind to govern thy people, that I may discern between good and evil.   (3:9)

The meaning of Solomon's vision is defined by the famous case of the dispute beween the two harlots over who is the rightful mother of a child. Solomon demonstrated great wisdom when he decided the child was to be cut in half in order that each mother might receive a fair share. The true mother protested, revealing her love for the child, whereupon Solomon ruled justly. In the past, when an issue such as this arose, the Urim and Thummim were consulted. But in this passage, Solomon's great wisdom is sufficient to solve the case. The king presides as a judge with ability to render divine justice, a role formerly reserved for priests. The king's wisdom replaced the Urim and Thummim. Although the administration of justice belongs ultimately to God, the king in this judicial function represented God.[37]

The same view of the just king is reflected in the words of the Tekoite widow when she refers to David as an angel of God discerning good and evil (II Sam. 14:17) and knowing all things that are on the earth (14:20). Mephibosheth uses the same phrase when he asks for mercy from David (19:27 [H 19:28]).

Probably during the reigns of David and Solomon the ideal of the just king entered Hebrew faith. There the figure found a receptive response; the prototype became the basis for the messianic hope of the postexilic period. The just-king ideal is most clearly expressed in the royal psalms, proverbs, and a few prophetic oracles.[38] Comparing the Hebrew view of the just king with similar idealized figures in Egypt, Mesopotamia, and Canaan (see Chapter 2) reveals how greatly the Hebrew people were influenced by their neighbors.

No description of the just king is better than that found in Psalm 72.

> Give the king thy justice, O God,
>   and thy righteousness to the royal son!
> May he judge thy people with righteousness,
>   and thy poor with justice!

> May he defend the cause of the poor of
>   the people,
> give deliverance to the needy,
> and crush the oppressor!
>
> (vv. 1–4)

> For he delivers the needy when he calls,
>   the poor and him who has no helper.
> He has pity on the weak and the needy,
>   and saves the lives of the needy
> From oppression and violence he redeems
>   their life;
> and precious is their blood in his sight.
>
> (vv. 12–14)

The king's concern for the needy and the poor assures the nation's fertility and prosperity (vv. 6–16).[39]

The Book of Proverbs contains similar references to the just king.

> By justice a king gives stability to the land,
>   but one who exacts gifts ruins it.
>
> (Prov. 29:4)

> It is an abomination to kings to do evil,
>   for the throne is established by righteousness.
> Righteous lips are the delight of a king,
>   and he loves him who speaks what is right.
>
> (Prov. 16:12–13)

Most familiar are certain prophetic oracles that define the just king, especially those contained in the Book of Isaiah.

> For to us a child is born,
>   to us a son is given;
> and the government will be upon his shoulder,
>   and his name will be called
> "Wonderful Counselor, Mighty God,
>   Everlasting Father, Prince of Peace."
> Of the increase of his government and of peace
>   there will be no end,
> Upon the throne of David, and over his kingdom,
>   to establish it, and to uphold it

with justice and with righteousness from
   this time forth and for evermore.
The zeal of the Lord of hosts will do this.
                    (Isa. 9:6–7 [H 9:5–6])

He shall not judge by what his eyes see,
   or decide by what his ears hear;
but with righteousness he shall judge the poor,
   and decide with equity for the meek of the earth;
                    (Isa. 11:3b–4a)

As we have seen, both David and Solomon established a bureaucracy in Jerusalem and appointed officials to cities and fortresses to
administer the emerging state.[40] One of the functions of these officials
was to assure that both the cultic and the civil laws were upheld.
They undoubtedly used the military to see that laws were enforced.
The exact relationship between these monarchical appointees and the
local clan-tribal legal system probably varied with each particular
situation, but certainly the center of judicial power was shifting to
the monarchy.

Before closing this section on royal judicial process during the
early monarchy, we should consider two other examples of monarchical legal procedures. The first concerns land tenure. Sprinkled
throughout the narrative are references to the power of the crown
over certain land estates. Both Saul and David used royal land grants
to secure loyal subjects (I Sam. 22:7–8). When David fell heir to
Saul's land (II Sam. 12:8), he befriended the invalid son of Jonathan,
Mephibosheth, by giving him Saul's estates (9:1–3).[41] When Mephibosheth later was accused by his servant Ziba of disloyalty toward
David during Absalom's rebellion (16:1–4), David tried unsuccessfully to determine who was telling the truth, then finally divided the
land between them (19:24–30).

The second use of royal judicial powers is the declaration of
amnesty at the time of coronation. This occurs frequently in the
text. When Saul was made king at Gilgal (I Sam. 11:12–15), he
granted amnesty to all those who opposed his elevation to the
throne.[42] David became king and spared Mephibosheth (9:1–13).
After Absalom's rebellion and David's return to the throne at Jerusalem, David spared Shimei who had cursed him (19:16–23).
Solomon was crowned king; he then exiled the priest Abiathar,
who Solomon believed deserved to die for supporting Adonijah

(I Kings 2:26–27).[43] Amnesty at coronation was a king's prerogative during this period.

The Monarchical Judicial System
in the Northern Kingdom of Israel

There is little data available concerning the royal court system in the northern Kingdom of Israel. The case of the unknown prophet (I Kings 20:35–43) has characteristics of the literary construction found in the parable of Nathan and in the account of the widow of Tekoa. The case does demonstrate that, as in the days of the united monarchy, King Ahab of Israel was above the law. However, the prophetic story illustrating this maxim is not sufficiently similar to what Ahab had done to make the analogy valid.

The case of Naboth (21:1–19a),[44] on the other hand, does raise issues concerning the relationship of the local courts to the royal court. According to the account, Ahab desired Naboth's vineyard because of its proximity to the palace in Jezreel. The king offered either a better vineyard in exchange or, if Naboth desired, a fair price. This was an honest offer; Naboth could have complied with the request. Whitelam appropriately calls attention to Leviticus 25:29–31 that allows for the sale of family property in the city; village property is to be kept for one's descendants.[45] Naboth decided to reject the king's offer, and Ahab considered the matter closed. But Ahab's wife Jezebel sent letters in the king's name and over the royal seal to the elders and nobles where Naboth lived. Naboth was charged with having cursed God and the king. The exact nature of the accusation is obscure but was sufficiently grave to warrant a public fast to allow the community to dissociate itself from the offense. In the public trial, two witnesses seated opposite Naboth brought false charges against him. Naboth was convicted and executed by stoning outside the city; the punishment further indicates the gravity of the charge. The king's subsequent confiscation of Naboth's vineyard attests that the charge was probably some form of treason.

Although the charges were false and the entire sordid episode reveals the crown's misuse of power, there is no indication that the trial was conducted illegally. That is, had the charges been true, the conviction, sentencing, execution, and confiscation of property by the crown would be legal.[46] The account indicates that the local courts on occasion were subordinate to monarchical judicial authority. Although the trial took place within Naboth's local community, the monarchy exerted influence over the outcome.[47]

The placement of Naboth's trial within the Elijah cycle implies that the king was above the law. It took a prophet, speaking for God, to condemn the conduct of Jezebel and Ahab. They were subject to divine justice, but apparently not to civil justice.

The case of cannibalism in Samaria (II Kings 6:24–31) reveals little about monarchical judicial procedures.[48] Although some of the language is judicial, the material is legendary and the case is bizarre. There is no trial and no decision. But the case may show that the king had jurisdiction over local issues in Samaria, perhaps because the crown purchased the land at the outset (I Kings 16:24).[49]

The suit of the Shunammite woman (II Kings 8:1–6) deals mainly with land tenure. During a time of famine, she left the country and spent seven years in Philistia. Upon returning, she discovered her land had become part of the royal estates.[50] She appealed to the king to restore to her the title to the land. The king, apparently on his own, made the decision in her favor and appointed an official to see that she received fair compensation for what had been produced on the land while she was abroad. If the case is historically accurate, it reveals a very humane judicial system operating, probably during the reign of Jehoram (849–843/2), son of Ahab.

The Judicial Reforms of Jehoshaphat

By far the most important judicial reforms occurred in Judah during the reign of Jehoshaphat (873–849). According to II Chronicles 17:1– 9, Jehoshaphat fortified cities and established garrisons in Judah and in parts of occupied Ephraim. He then began a religious reform by sending out princes (*śārîm*), Levites, and priests to instruct the people on religious matters from the "book of the law." According to 19:4– 11, Jehoshaphat was also a judicial reformer. In fortified cities he placed appointed judges to administer the law (vv. 5–7). In Jerusalem he established a court and appointed Levites, priests, and heads of families to decide disputed cases (vv. 8–11). These reforms are extremely important in understanding Amos' call for social justice in Israel. They must be examined carefully.

The first issue we must address is whether the chronicler's references are historically accurate. It is fairly easy to see why Jehoshaphat's reforms are not included within the deuteronomistic history of Israel. The deuteronomistic historians were antinorthern in their interpretation of the history of Israel and Judah; they regarded Ahab as one of the worst kings of the north. Because Jehoshaphat was a known ally of Ahab (I Kings 22), praising Jehoshaphat for his legal reforms did not suit the deuteronomistic historians' purpose. Yet the historians did commend Jehoshaphat for walking in the way of his father Asa, "doing what was right in the sight of the LORD" (22:43a), which included doing away with male cult prostitutes (22:46). Did the deuteronomists know of the reforms? Scholars have long recognized that Deuteronomy 1:9–19; 16:18–20; and 17:8–13 allude to Jehoshaphat's reforms (see also Exod. 18:21, 25).[51] Though attributed to Moses, the references to the appointment of judges in settled towns (Deut. 16:18; 17:8) and to the referral of difficult cases to "the place which the LORD your God will choose" (17:8) do not reflect the Mosaic period. The deuteronomists attributed these reforms to Moses, because they believed that all laws come from Yahweh through Moses and that nothing was to be added to them (4:1–2).

On the other hand, we know that the chronicler wished to legitimize postexilic institutions by tracing them back to the Davidic dynasty. If these judicial reforms are postexilic, as some scholars argue,[52] it would have better suited the chronicler's purpose to relate them directly to the work of David. There is no reason to attribute them to Jehoshaphat unless that is historically accurate. Further evidence in support of the historicity of Judah's reforms are the eighth-century prophetic attacks on the corrupt Judahite judges and leaders (see Isa. 1:21–26; 3:2–3, 14; Mic. 3:1–2, 9–12; 7:3). We have no knowledge that these officials existed before the reforms of Jehoshaphat.

The nature of Jehoshaphat's reforms are threefold. First, he continued the major religious reforms of his father Asa (II Chron. 19:4), who sought to concentrate worship in Jerusalem where it could be properly regulated (I Kings 15:9–14; II Chron. 14:1–5 [H 13:23–14:4]).

Second, he appointed judges in all the fortified cities of Judah and instructed them to provide justice in the provinces (II Chron. 19:5–7). Most likely, Jehoshaphat was using the existing military organization to establish royal judicial authorities at strategically placed garrisons.[53] Deuteronomy 16:18 instructs that these judges are to be placed "in all your towns." The latter order is either an attempt to make the reforms applicable to all the land or reflects Josiah's later effort to extend Jehoshaphat's reforms. In any case, Jehoshaphat obviously used the military organization as a natural base for establishing a strong fiscal and judicial administration in the provinces. These royal courts probably had four different types of officials within them—heads, elders, judges, and officers (see Deut. 29:10 [H 29:9]; Josh. 8:33; 23:2; 24:1). The local village court apparently had only two types of officials—heads of families and elders. What was the relationship between the military-backed, royal court system of the fortified cities and the local, clan–village judiciary? Perhaps the fortified-city courts heard cases referred to them alone by the local village courts. More likely, however, the monarchical city court replaced clan-tribal courts to become the chief judicial court in the land. A. Phillips argues that these royal judicial courts used the Book of the Covenant, which called for humaneness and righteousness, as a handbook on justice for the Davidic state.[54]

Third, Jehoshaphat established a court in Jerusalem (II Chron. 19:8–11). Despite textual problems with v. 8, we must understand that the Jerusalem court is not a court of appeals, but rather a court to which local officials look for guidance in making judicial decisions. The court was to instruct the local judges how to rule justly and fairly. The reference to a two-court system in Jerusalem, a civil and sacral court, probably reflects the postexilic period of governor Zerubbabel and chief-priest Joshua.[55] The premonarchical period contains no sharp distinction between these two realms in law.

Amos and Jehoshaphat's Reforms

What motivated Amos to deliver his message of social justice in Israel? In recent years, social anthropologists claim the prophetic ethic must be related to the egalitarian revolution, which they believe occurred in Ancient Israel during the premonarchical period. At that time, they argue, Israel was formed from a rebellious agriculturalist and pastoralist Canaanite people who opposed the centralized structure of their city-states. This revolution against repressive overlords took the form of a "retribalization" movement toward the simple politics of the village-based life of the clan–tribe. Social justice was established through clan-tribal councils meeting at the city gate. Possession of land was patrimonial; families and clans controlled land granted to them by Yahweh. These estates were passed from father to son, occasionally redistributed within the local clan.[56]

With the emergence of the monarchy in Israel, these scholars believe the clock turned back. The Hebrew people returned to the bondage they had once known under Canaanite overlords. The authority of the local clan-tribal court diminished, replaced by a royal court system. The ruling elite gradually acquired possession of the land through policies of agricultural land tenancy and confiscation. The peasants became disinherited and landless; the king became the major land owner, not Yahweh. Ownership of the land shifted from the patrimonial system to the prebendal system, in which the king granted estates to his officials in return for services rendered.[57]

Amos' message of social justice, social anthropologists argue, was a call for return to the egalitarian days of the premonarchical period when there was clan justice at the gate, where peasants could earn decent livings on their lands, and where fair prices were charged in the marketplace.[58]

This approach to prophetic ethics is not without its critics. As we have seen, an exclusive Canaanite peasant origin of the Hebrew people is questionable. A complex, multiple origin is far more likely.[59] The "peasant revolt" interpretation of the conquest and settlement period also depends heavily on analogies drawn from other cultures far removed in distance and time from Ancient Israel. Even its advocates confess its striking similarity to the peasant revolutions occurring during the twentieth-century c.e. in South America.[60] The temptation is strong to interpret the ancient situation in terms of modern events, but we should not forget that by the time of Amos approximately 400 years had passed since the so-called ideal age of the peasant revolt. Israel had experienced a monarchy for 250 years. While that monarchy was often repressive, the ideal was still the just king. The messianic passages within the prophetic oracles do not indicate that the prophets had given up hope that the next king would establish justice and righteousness.[61] Both Isaiah of Jerusalem and Micah, who were critical of how the officials appointed by the crown administered justice (see Isa. 1:21–26; 3:2–3, 14; Mic. 3:1–2, 9–12), still looked to the dynasty of David as the true guarantor of social justice (see Isa. 9:2–7 [H 9:1–6]; 11:1–9; Mic. 5:2–4 [H 5:1–3]). Even if these passages are exilic or postexilic in their present form,[62] they still reveal that the people looked to the monarchy for their deliverance from oppression. Furthermore, the egalitarian approach to Amos' message of social justice does not explain adequately why he went north to prophesy rather than remain with his own people in the south.[63]

Placing Amos within a nationalistic setting provides a more likely base for understanding his critique of social injustices in the north. It is reasonably certain that the royal judicial system established by Jehoshaphat included Tekoa as one of its administrative centers. II Chronicles 11:5–12 lists fifteen cities that Rehoboam fortified in Judah and Benjamin for the defense of the nation. One of these cities was Tekoa (v. 5). Because Rehoboam placed commanders in these cities, it is not unlikely that they helped him administer the state.[64] One of Jehoshaphat's legal reforms was to appoint judges in all the fortified cities of Judah to see that justice was established throughout the land (II Chron. 19:4–7); there is some indication that Tekoa was one of these cities. II Chronicles 20:1–30 narrates the story of Jehoshaphat's victory over the Moabites, Ammonites, and the men of Mount Seir. The story is primarily used by the chronicler to show the faith of Jehoshaphat, which resulted in his deliverance when the foreign armies destroyed one another. The story is unique to the chronicler, and we have no way of assessing its historicity.[65] But the battle purportedly occurred in "the wilderness of Tekoa" (v. 20), indicating the region was known for warfare. Apparently this remained the case throughout the existence of Judah, for Tekoa is mentioned as a defense city as late as the time of Jeremiah.

> Flee for safety, O people of Benjamin,
>   from the midst of Jerusalem!
> Blow the trumpet in Tekoa,
>   and raise a signal on Beth-haccherem;
> for evil looms out of the north,
>   and great destruction.[66]
>   (Jer. 6:1)

It is reasonably certain that Amos had the opportunity to observe the administration of justice in the royal court that convened in his native city.

Yet Amos gave no hint in his oracles what he thought of Judah's judicial system. Here he is in sharp contrast with Isaiah of Jerusalem and Micah. Isaiah (ca. 742–701), prophesying in Jerusalem, condemned the corruption that he witnessed within the leadership of that judicial system.[67] Micah (ca. 701), like Amos, came from a small village in Judah called Moresheth of Gath, located twenty-five miles southwest of Jerusalem (Mic. 1:4). In oracles spoken from Jerusalem (Mic. 3:1–2, 9–12), he condemned those officials who, acting as judges in the provincial court, used their powers to acquire wealth for themselves at the expense of the poor.[68] The authentic Amos oracles concerning social injustices are addressed only to Israel. Once again Amos reveals his nationalistic leanings toward Judah with its legitimate Davidic dynasty authorized by God to establish a judicial system in which justice is fairly administered.

Social Justice in Amos

Amos did not hesitate to deliver a scathing attack against those persons in Israel whose behavior he found reprehensible. He named names and pinpointed the precise acts he considered sinful. To avoid allowing this discussion of social justice in Amos to degenerate into vague generalizations, I have chosen to organize the material under three specific headings. They are not mutually exclusive.

Oppression of the Poor and the Needy

In the book's first passage focusing on a crime of Israel, Amos condemned the oppression of the poor and the needy.[69]

> Thus says the LORD:
> "For three crimes of Israel, and for four,
>   I will not cause it to return;
> because they sell the innocent for the silver,
>   and the needy for a pair of sandals—
> they that trample the head of the poor into
>   the dust of the earth,
>   and turn aside the way of the oppressed . . . "
>   (2:6–7a; author's translation)

This is a favorite topic for Amos. Three additional passages link the poor and the needy, centering the reader's attention on their plight (4:1; 5:11–12; 8:4–7).

In 2:6b–7a Amos pleads the case of the needy (*'ebyôn*) and the poor (*dallîm*). These were the lowly, weak, helpless people, the disinherited of the land.[70] They were the small farmers who no longer owned their own lands; they had become tenant farmers who paid rental fees to the absentee landowners who dwelled in the cities. They had no hope of recovering their ancestral lands, the key to being free and independent farmers.

Both couplets indicate the poor and the needy are being exploited illegally. In 2:6b the parallel term for needy is innocent (*ṣaddîq*), persons who are guiltless before the law. The innocent are sold into debt-slavery for "the" silver. The Masoretic text contains the definite article, probably designating a particular amount owed the creditor.[71] How were they innocent? Perhaps they were sold to settle the parent's debt (II Kings 4:1), or perhaps the debt they incurred was through no fault of their own. It is more likely, however, that the creditor had illegally forced them into debt-slavery. Similarly, the needy were sold "for a pair of sandals." This may mean their debts were trivial, only the value of a pair of sandals.[72] However, this is a legal phrase to indicate a transfer of land; the sandals are placed in the hand of the new owner (Ruth 4:7–12; Deut. 25:7–10).[73] Equating the needy with the innocent means the former group probably lost possession of their property through some illegal conduct by a creditor. In 2:7a the poor are in synonymous parallelism with the oppressed (*'ănāwîm*). "To turn aside the way of the oppressed" is used as equivalent to perverting justice in the gate (see Prov. 17:23).[74] Amos was referring to injustices committed by the clan elders when trying cases in the city gate. Both passages, taken together, infer the needy and the poor were innocent parties in a legal process being used to exploit them.

In 4:1 Amos contrasted the poor and the needy with the wealthy. Although the verse may allude to the *marzēaḥ* festival,[75] the thrust of the verse is an attack by Amos on the wives of wealthy merchants, landowners, and court officials who lived in luxury and were indifferent to the plight of those forced to live in poverty. Yet the tone of the passage is so harsh and the punishment so severe (4:2–3) that Amos possibly had in mind an illegal acquisition of wealth by the upper classes at the expense of the lower classes. This is certainly the case in 5:12 in which the clan elders had allowed bribes to influence their voting in the courts at the city gates so that decisions went against the poor and the needy.

Wolff argues that 8:4–7 is a complete judgment oracle, which he attributes to the Amos School ca. 735 B.C.E. He finds the broad indictment (vv. 4–6) and the attitude toward festival life (v. 5) uncharacteristic of Amos, while v. 6 is an artificial statement of the authentic Amos saying of 2:6b–7a.[76] Other critics accept the unit as authentic Amos.[77] Although the merchants do not break the sabbath law of rest, they do lack the proper attitude of using holy days to honor God. The main thrust of the verses, however, condemns illegal practices to the marketplace. The ephah is a dry measure, about two-thirds of a bushel, used to measure grain to be sold. Merchants cheated the customer by reducing the size of the ephah. Or the merchants might mix chaff with the wheat, thereby increasing their profit. In Amos' day, before the use of minted coins, the shekel was a weight for determining the amount of silver needed to purchase a commodity. A heavy weight increased the cost of the item. In addition, the cross beam in the balance could be bent slightly so as to cheat the customer. All these methods resulted in the economic enslavement of the poor.

Judgment Oracles Attack the Upper Classes

Amos' strongest judgment oracles attack the wealthy upper classes in Israel. Arvid S. Kapelrud argues one of the major contributions of Amos to Hebrew religion is his equation of poverty with righteousness (2:6; 5:12b) and, conversely, wealth with sinfulness.[78] In preexilic Hebrew religion, a customary precept was that God rewards those who are righteous with both spiritual and material blessings in this life. Prosperity was outward proof of a person's piety. This is most clearly reflected in prudential wisdom literature (e.g., Prov. 3:5–10; 4:10, 13; 13:18, 21; cf. the arguments of Job's three friends). Amos' insistence that it was the poor and the needy who were innocent must have shocked those who heard his message. Amos had not one kind word to say about those who were rich, and he had not one critical word to say about those who were poor. His scathing condemnation of the upper classes and his compassion for the poor have led some scholars to designate him as the patron prophet of liberation theology.[79]

One of the means by which the upper classes acquired their wealth was through land grants from the monarchy. In return for loyalty and service, they received estates the crown acquired through conquest and confiscation. The wealthy could expand their landholdings by foreclosing on the poor who were unable to pay their debts. In addition to land possession, there were also profits to be made as merchants within growing cities.

Amos mounted an attack against the upper classes for three specific sins. First, the upper classes were guilty of living in luxury while the lower classes suffered in poverty. Does this mean that Amos would not have condemned the wealthy aristocrats had they used their resources to relieve the needs of the poor? This we do not know. What we do know is that all references in the Book of Amos condemn luxurious living as sinful.

A surprising number of details emerge from the text that reveal the life-style of the wealthy. Some possessed more than one house—having a summer home in the hill country where it was cool and a winter home in the valley where it was warm (3:15).[80] I Kings records that in a previous century King Ahab of Israel owned one palace in Jezreel (I Kings 21:1) and another in Samaria (21:18). While the poor lived in houses constructed with clay bricks that crumble easily (see Isa. 9:10 [H 9:9]), the wealthy possessed houses constructed with close-fitting cut stone (Amos 5:11).[81] The wealthy filled their homes with fine furniture decorated with ivory inlay (3:15; 6:4), some of which has been discovered in excavations at Samaria.[82] The woe-oracle in 6:1, 3–7 describes the life of leisure the upper classes enjoyed. They reclined on couches, ate tender meats such as young lambs and calves confined to fattening stalls, drank excessively from large bowls, rubbed their bodies with expensive oil, and sang noisy songs (6:4– 7).[83] They were a pampered group that had their slightest desires fulfilled. Amos likened upper-class wives to the well-tended cows of Bashan; just as the cows of Bashan had owners who provided them with lush fields for pasturage, so the aristocratic wives had husbands who supplied them with intoxicating drink (4:1).[84] Amos perceived that such lives of leisure led to moral corruption. With 2:6–8 an indictment of the wealthy upper classes, v. 7b then condemns an aristocratic father for gross sexual immorality. The son, who is presumably unmarried, has had sexual relations with a marriageable girl (*hannaʿărâ*). He is therefore obliged to marry her (see Deut. 22:28–29). The married father is guilty of having violated his future daughter-in-law (Lev. 18:15; 20:12).[85]

The second criticism Amos made against the upper classes was their lack of compassion for the poor. Amos saw how easily the lower classes lost their inherited lands when crop failures deprived them of means for earning a living. A law in the Book of the Covenant instructs a lender who has taken a neighbor's garment as security for a loan, to return it at night lest the poor suffer from the cold (Exod.

22:26–27; cf. Deut. 24:12–13). Amos condemned the lender not only for violating the law, but also for having the audacity to imagine God did not care.

> they lay themselves down beside every altar
> upon garments taken in pledge[86];
>
> (2:8a)

The insensitivity of the wealthy to the plight of the poor is particularly apparent in 5:11. Amos condemned the landowner for insisting on rent payment in wheat that the tenant farmer could not afford and then for using the income to finance the building of a fine house and the planting of a well-designed vineyard.

Amos' third criticism of the wealthy was for their blatant dishonesty both in the judicial courts and in the marketplaces. The clan elders who were to establish justice in the gates broke the law by taking bribes to decide against the poor; the poor were then found guilty, although they were innocent (2:6b–7a; 5:12). The merchants gave short measure, used false balances, and mixed chaff with the wheat (8:4–7). The state was apparently unable to administer the law effectively.

Call for Justice and Righteousness

Amos' essential ethical message is a call for justice and righteousness in Israel. "Justice" and "righteousness" form a word-pair of central importance in understanding Amos' attack on the rich and his compassion for the poor. They are paired in three passages, and justice appears alone in one reference.

> O you who turn justice to wormwood,[87]
> and cast down righteousness to the earth!
>
> (5:7)

> But you have turned justice into poison
> and the fruit of righteousness into wormwood—
>
> (6:12b)

> But let justice roll down like waters,
> and righteousness like an ever-flowing stream.
>
> (5:24)

> Hate evil, and love good,
> and establish justice in the gate.[88]
>
> (5:15a)

Justice (*mišpāt*) and righteousness (*ṣĕdāqâ*) are rich theological terms used to characterize Yahweh and to describe the covenant relationship between him and his people.[89] A current tendency is to deemphasize their judicial aspects in order to avoid a "legalistic" approach to Hebrew ethics. This can unfortunately result in robbing the two of their strong moral content. To use the terms of contemporary liberation theology, the emphasis is on *orthodoxy* (right belief) rather than *orthopraxis* (right conduct). This is not the emphasis of Amos. The setting for justice in Amos is the legal court system. For Amos, justice meant the judicial process that established what and who is right, thereby preserving the well-being of the community. Justice prevails when right legal decisions that protect the weak and the needy are made in the courts by responsible persons. Righteousness designates the conduct of persons who are themselves legally innocent (*ṣaddîq*) in standing up for those who are unjustly accused (*ṣaddîq*).

In the Book of Amos, these legal decisions are being made in the gate where local court cases are heard. There are three references to the court meeting in the gate (5:10, 12, 15). Amos 5:7, 10–11 especially enlarge the description.[90]

> O you who turn justice to wormwood,
> and cast down righteousness to the earth!
> They hate him who reproves in the gate,
> and they abhor him who speaks the truth.
> Therefore because you trample upon the poor
> and take from him exactions of wheat,
> you have built houses of hewn stone,
> but you shall not dwell in them;
> You have planted pleasant vineyards,
> but you shall not drink their wine.

This is a woe-oracle that condemns those who pervert justice in the courts and those who hate anyone who defends the cause of the poor. These groups will not benefit from their manipulation of the court system. They will neither dwell in their fine homes nor drink the wine from their well-designed vineyards. Amos 5:12 is an indictment of those who take bribes and vote against the falsely accused needy.

Amos 5:13 is an exhortation to establish justice at the court meeting in the gate.

In 5:15a and 24, Amos exhorted those in the north to establish a just legal system. What did he have in mind? According to Wolff, the pairing of justice and righteousness in Amos comes from a wisdom tradition.[91] The latter was further embodied in a clan legal system operative within the courts that met in the city gates. Wolff argues that Amos probably first encountered this view of justice and righteousness at the court in Tekoa. In a similar vein, Gottwald writes:

> The substance of the socioethical laws of Israel was known to him even if he had never seen written laws. These laws were more or less faithfully practiced before his eyes in villages like Tekoa, as they were grotesquely ignored and overridden in the governing circle of Israel by the very people who were loudest in their praise of Yahweh.[92]

Yet as we have seen, Tekoa quite likely had an appointed official who presided over a royal legal court. Ever since the time of Solomon, the king was regarded as God's appointed representative to establish justice and righteousness (Ps. 72:1–2; II Sam. 8:15), to defend the poor and the needy. It was Jehoshaphat who established this royal court system in all his fortified cities. This legal system had been in place at Tekoa for approximately 100 years. Could not Amos have had this royal legal system in mind rather than the clan-tribal court system?

Did Israel possess a royal court system at the time of Amos? Hosea was a northern prophet shortly after Amos (ca. 735). He held the priests, clan chieftains, and the *royal court* ("house of the king") responsible for justice in the land (Hos. 5:1).[93] Scholars lack exact data on the system, but it may have been similar to that found in Judah, where the royal court either heard cases referred to it by the local courts or gradually replaced the clan-tribal courts to become the main court system in the land. Although most references to legal injustice in the Book of Amos apply to local courts meeting in the gates, Amaziah, the priest at Bethel, clearly had certain royal administrative duties.[94] He was definitely an official of the crown, appointed by the king to function at the temple of Bethel, designated "the king's sanctuary" (7:13). He was responsible for reporting serious crimes to the king (7:10). He also possessed land of his own (7:17), by the context, probably a grant from the crown in return for loyal service.[95] Are there indications that Amos was tried in a northern royal court meeting at Bethel? Although details of court procedures are lacking, Amos was possibly charged with conspiracy (7:10) and sentenced to return to Judah and never again to set foot in Bethel (7:12). No witnesses would have been necessary, because Amos readily admitted that he had prophesied that the king would be killed (7:11) and that the nation, including Amaziah, would be exiled (7:17).

It may be significant that 5:15a and 24, two exhortations to establish justice, are located within cultically related material. We saw earlier that two major functions of the kings in the Ancient Near East (Chapter 2) and in Israel and Judah (Chapter 3) were to establish and maintain temples to honor the gods who placed kings on their thrones, and to administer justice in the land. Temple and law court were closely related, as shown by the Amaziah incident. If Amos did condemn all northern cult centers as unacceptable worship sites and if he did call Israel to seek God only in Jerusalem (Chapter 5), then he quite possibly expected that the court system established by the Davidic kings offered hope to the poor and the needy. Both Isaiah of Jerusalem and Micah report that this system failed to establish true justice, but nonetheless they prophesied that God would some day raise up a just ruler who would establish justice and righteousness (Isa. 9:2–7 [H 9:1–6]; 11:1–5; Mic. 5:2–4 [H 5:1–3]). Likewise Amos looked forward to that day when the two kingdoms would be one under the just rule of a Davidic king (Amos 9:9–12, esp. v. 10). We shall explore the details of this hope in the final chapter of this book.

7

Amos and the Call for Repentance

THE NONSPECIALIST in the study of Hebrew prophecy may be unaware of recent scholarship into the subject of the prophetic call for repentance. The general reader of Hebrew Scriptures assumes that the prophets called Israel and Judah to repent in order to escape destruction and to assure redemption. However, since the publication in 1951 of Wolff's article, "Das Thema 'Umkehr' in der alttestamentlichen Prophetie,"[1] there has been a tendency to minimize the place of repentance in the message of the preexilic prophets. Ten years later, with the publication of "Das Kerygma des deuteronomistischen Geschichtswerks,"[2] Wolff maintains that the unheeded prophetic call to repentance was the essential theme of both the deuteronomist and the deuteronomistic historians. This deuteronomistic interpretation of the role of the prophet was sufficiently influential that all preexilic prophets presumedly called God's people to repent in order to escape destruction. But a form-critical approach to the oracles of the preexilic prophets reveals that repentance was not the essential theme. As a consequence, students of preexilic Hebrew prophecy began to view all judgment oracles as announcements of inescapable doom, and to regard all salvation oracles as declarations of God's unmerited grace following a period of destruction. The message of repentance in these prophetic works is either subordinated to the judgment and salvation themes or attributed to later redactors.

In his book *Amos among the Prophets*, R. B. Coote attributes the A stage of the Book of Amos to the prophet Amos. The A stage, he writes, consists of forty verses that proclaim the deportation of

139

the oppressive ruling elite of Israel and the deliverance of the powerless peasants. There is no call for repentance. Stage B was produced by the Bethel editor (a member of Josiah's court); he reworked stage A in order to call all people of both the north and the south to worship God sincerely and to take up the task of justice if they were to escape destruction. Because Israel did not repent, it was destroyed. This should serve as a warning to the B-stage readers of what will befall them if they refuse to change.

However, most surveys of Hebrew prophecy continue to present what A. Vanlier Hunter labels "the majority opinion."[3] The highly esteemed Jewish scholar Abraham Heschel writes:

> Indeed, every prediction of disaster is in itself an exhortation to repentance. The prophet is sent not only to upbraid, but also to "strengthen the weak hands and make firm the feeble knees" (Isa. 35:3). Almost every prophet brings consolation, promise, and the hope of reconciliation along with censure and castigation. He begins with *a message of doom*; he concludes with *a message of hope*.[4]

A footnote indicates that Heschel is aware of the alternate position, but accepts the standard approach:

> Some modern scholars maintain that the preexilic prophets had no message except one of doom, that true prophecy is essentially prophecy of woe. Yet such a view can be maintained only by declaring, often on insufficient grounds, that numerous passages are interpolations.[5]

C. F. Whitley summarizes concisely "the majority opinion":

> But while certain passages point to a tension in the minds of the prophets whether they should denounce the people or call them to repentance, it is possible to discern in their teachings a pattern in which they at first entertained the hope of penitent Israel turning from her sins, but on her persistent refusal to heed their warnings they represented Yahweh's judgment as issuing in final and irrevocable doom.[6]

It is not difficult to demonstrate that this view of prophecy is found in both Deuteronomy and the deuteronomistic history of Israel. The normative passage defining the nature of Hebrew prophecy in

Deuteronomy is 18:9–22. In vv. 9–14 the Hebrew people are commanded to put away the practices of the soothsayer, augur, sorcerer, charmer, medium, wizard, and necromancer. God will communicate not through these offices, but through a prophet like Moses. The cardinal text is v. 18. God addresses Moses:

> I will raise up for them a prophet like you from among their brethren, and I will put my words in his mouth, and he shall speak to them all that I command him.

God will raise his prophet to instruct the people as Moses had instructed them on how to conduct their lives. In distinguishing between true and false prophets, however, the deuteronomistic historians proceed to disclose the prophet as a predictor of future events.

> When a prophet speaks in the name of the LORD, if the word does not come to pass or come true, that is a word which the LORD has not spoken. (Deut. 18:22)

The passage is ambiguous about whether the prophet is to give commandments and instructions on how to live, whether he is to predict what God is about to do, or both.

This theme of prophetic prediction and fulfillment is found throughout the deuteronomistic materials. The most obvious example is an unnamed prophet's prediction, delivered to Jeroboam I at Bethel, that some day a king named Josiah would desecrate the altar by burning human bones upon it and by destroying it (I Kings 13:1–10). II Kings 23:15–18 contains the fulfillment of the prophecy:

> Moreover the altar at Bethel, the high place erected by Jeroboam the son of Nebat, who made Israel to sin, that altar with the high places he pulled down and he broke in pieces its stones, crushing them to dust; also he burned the Asherah. And as Josiah turned, he saw the tombs there on the mount; and he sent and took the bones out of the tombs, and burned them upon the altar, and defiled it, according to the word of the LORD which the man of God proclaimed, who had predicted these things.

With this emphasis on prediction and fulfillment, the deuteronomistic historians were creating a view of prophecy that would help them understand the fall of Israel in 722/1 B.C.E. and the fall of Judah in 587. The Hebrew people by tradition understood that the prophets were persons who called God's people to turn about and be saved. They would be destroyed only when they refused to repent. When Solomon dedicated the temple (I Kings 8:33–55), he presented this call to repentance: If the people turn again and acknowledge God, they will be saved; if they rebel, they will be destroyed and exiled. They should have no complaints; they were forewarned. It is no surprise, therefore, that the deuteronomistic historians' interpretation of the fall of Israel reads:

> Yet the LORD warned Israel and Judah by every product and every seer, saying, "Turn from your evil ways and keep my commandments and my statutes, in accordance with all the law which I commanded your fathers, and which I sent to you by my servant the prophets." But they did not listen, but were stubborn, as their fathers had been, who did not believe in the LORD their God. ... Therefore, the LORD was very angry with Israel, and removed them out of his sight, none was left but the tribe of Judah only.   (II Kings 17:13–14, 18)

Hunter makes the telling point that, with the exception of Elijah's admonition in I Kings 18:21, the deuteronomistic history of Israel contains no quotations of prophetic calls for repentance? Only in the deuteronomists' own composition do such calls occur. Hunter argues that if the prophets had urged Israel and Judah to repent, the deuteronomists surely would have included the supporting evidence within their history. The best explanation of their failure to cite such examples is that the repentance theme was a deuteronomistic invention to explain why both the northern and southern kingdoms were destroyed.[7] This is what Hunter calls "the minority opinion," the position that the preexilic prophets preached inescapable doom because Israel and Judah had failed to obey God's will. It was the twelfth hour; the end had come.

Hunter's comparison of Jeremiah's words with the deuteronomistic biographical narrative in the Book of Jeremiah strongly supports "the minority opinion."[8] Jeremiah 26 contains the deuteronomistic version of the "temple address." The response of the people to Jeremiah shows what they heard was a judgment oracle:

> Why have you prophesied in the name of the LORD, saying, "This house shall be like Shiloh, and this city shall be desolate without inhabitants?" (v. 9)

The narrative account transforms this doom oracle into an exhortation
for repentance. The style is decidedly deuteronomistic:

> You shall say to them, "Thus says the LORD: If you will not listen to
> me, to walk in my law which I have set before you, and to heed the
> words of my servants the prophets whom I send to you urgently,
> though you have not heeded, then I will make this house like Shiloh,
> and I will make this city a curse for all the nations of the earth.
>   (vv. 4–6)

> Now, therefore amend your ways and your doings, and obey the voice
> of the LORD your God, and the LORD will repent of the evil which he
> has pronounced against you.   (v. 13)

What the people heard was an unconditional oracle of doom: the
temple and the city will be destroyed. But the deuteronomists represented Jeremiah as one who believed the people could escape destruction if they only repented.

What motivated the deuteronomists to transform doom oracles
into exhortations to repent? The deuteronomistic history of Israel
took final form during the Babylonian exile (587–538). According to
Wolff, the intention of the entire history is not only to explain why
Israel and Judah fell (they failed to repent), but also to encourage
the exiles to turn to God so that they might receive his blessings.[9]
Nowhere is this latter purpose better expressed than in Moses' speech
in Deuteronomy 30:1–10:

> And when all these things come upon you, the blessing and the curse,
> which I have set before you; and you call them to mind among all the
> nations where the LORD your God has driven you, and return to the
> LORD your God, you and your children, and obey his voice in all that
> I command you this day, with all your heart and with all your soul;
> then the LORD your God will restore your fortunes; and have com-
> passion upon you, and he will gather you again from all peoples where
> the LORD your God has scattered you. If your outcasts are in the
> uppermost parts of heaven, from there the LORD your God will gather
> you, and from there he will fetch you; and the LORD your God will
> bring you into the land which your fathers possessed, that you may
> possess it; and he will make you more prosperous and numerous than
> your fathers. And the LORD your God will circumcise your heart and
> the heart of your offspring, so that you will love the LORD your God
> with all your heart and with all your soul, that you may live. And the

LORD your God will put all these curses upon your foes and enemies who persecuted you. And you shall again obey the voice of the LORD, and keep all his commandments which I command you this day. The LORD your God will make you abundantly prosperous in all the work of your hand, in the fruit of your cattle, and in the fruit of your ground; for the LORD will again take delight in prospering you, as he took delight in your fathers, if you obey the voice of the LORD your God, to keep his commandments and his statutes which are written in his book of the law, if you turn to the LORD your God with all your heart and with all your soul.

Similarly, Solomon's prayer at the dedication of the temple conveys this same interpretation of the exile (see I Kings 8:46–53)—that if the people repent, God will save them.

The two positions on the essential message of the preexilic prophets, which Hunter designates "the majority opinion" and "the minority opinion," are supported or denied by certain key passages in Amos. We shall examine various interpretations of these passages in some detail. Two questions are before us: (1) Is there any evidence that the prophet called Israel to repent in order to escape destruction? (2) Did Amos proclaim inevitable destruction because of the nation's past crimes?[10]

Amos 4:4–13

It has been customary to divide this passage into three independent units (vv. 4–5, 6–12, 13). Some scholars, however, consider it a unified speech form. Certainly the redactional approach provokes questions of why the compilers arranged the material in this order.

The key to interpreting the passage is God's words to Israel in v. 12:

> Therefore thus I will do to you, O Israel;
> because I will do this to you
> prepare to meet your God, O Israel!

Is this an eleventh-hour call for repentance or the sounding of a death knell?

The majority opinion interprets the phrase "prepare to meet your God, O Israel" as a call to return to God and be saved. God is offering Israel one last chance to repent and avoid destruction. John Marsh writes:

> This is not a threat or judgment, but a call to repentance. As if Amos had said, "Don't let it have to be said again, but ye have not returned to me, saith the Lord, but instead prepare to meet God. Repent; turn to him in a new righteousness and a sincere worship."[11]

Although Harper senses the ominous nature of the threat, he still argues that "every prediction of disaster was in itself an exhortation to repentance, in order that, if possible, the disaster might be averted."[12]

W. Brueggemann proposes that the entire unit is both a liturgical formula of preparation for covenant renewal, and a summons to prepare for holy war.[13] The unit contains both a promise and a threat. It begins with a reference to a false covenant formed at northern shrines:

> Come to Bethel, and rebel
> to Gilgal, and rebel even more.
> (v. 4)

There follow words of judgment on the broken covenant (e.g., "I gave you cleanness of teeth in all your cities" and "I smote you with blight and mildew"), each concluding with the refrain "yet you did not return (*wĕlōʾ šabtem*) to me." This refrain is repeated five times. The section closes with the phrase "prepare to meet your God, O Israel" (v. 12c). Finally come words of praise to the covenant God (v. 13), one of the three doxologies in the Book of Amos. Brueggemann argues that the words "prepare to meet your God, O Israel" have double meaning. They are, first of all, a liturgical summons to Israel to prepare itself through cultic acts for covenant renewal. Second, if Israel refuses to repent, it should be prepared through cultic sanctification for God's holy war conducted now against Israel rather than against other nations.[14] What is placed before Israel are the alternatives of blessing (covenant renewal) or curse (holy-war combat with God). The time for decision has come. Yahweh had punished past rebellions; he now gives Israel one last chance to turn about (*šûb*) and be saved.

Wolff attributes vv. 4–5 to Amos, but ascribes vv. 6–13 to an

addition at the time of Josiah by the Bethel editor who applied them
to the destruction of Bethel.[15] Wolff envisions the redactor delivering
the oracle at the destroyed site of Bethel. Pointing to the ruins, he
said:

> Therefore, *thus* I will do you, O Israel;
> because I will do *this* to you,
> prepare to meet your God, O Israel.
>
> (v. 12)

Wolff's position is that Amos preached total destruction, but that the
Bethel editor in Josiah's day called the people to repent.

George Ramsey, in his article on 4:12, suggests the entire unit of
vv. 6–12 as an ironic use of a *rîb* pattern.[16] A *rîb* oracle, he maintains,
normally recounts God's acts of mercy and then calls for repentance.
After reviewing God's acts of judgment, the section closes with the
words:

> prepare to call your gods, O Israel.
>
> (v. 12c)

Ramsey argues that the verse uses the first meaning of the infinitive
qārā', "to call or summon," not its second meaning, "to meet."
Further, he translates *'ĕlōheykā* as a plural reference to "gods," al-
luding to foreign deities, not Yahweh. The passage summons Israel
to call upon the foreign deities it now worships, but the text implies
these gods will not listen. The end is near; destruction is at hand.
Despite the attractiveness of this interpretation of v. 12c, Ramsey
has succumbed to a somewhat forced use of satire. And, as we considered earlier, the evidence is not strong that Amos thought Israel
was worshiping foreign deities.

On the other hand, A. V. Hunter rejects the repentance theme
held by a majority of scholars. After studying in detail the exhortation
passages in five postexilic prophets, he supports the minority opinion
that these verses are a doom oracle. His summary statement follows:

> The imperative clause "prepare to meet your God" in Am 4:12b is
> not an exhortation to repentance but an ironic call for readiness to
> encounter Yahweh who is coming to his people in judgment. Amos
> purposefully chooses wording from the cult that could have a double
> meaning. He proclaims that Israel should be ready to experience a
> theophany, or at least a cultic version of a theophany. The verbal

links with the Sinai tradition would lead the people to expect Yahweh's coming, however awe-striking it may be, to bring salvation. Amos, in one of his typical reversal moves, really intends the other verbal association, namely, that Yahweh's appearance, which will be real and not merely symbolized in the cult, will be as an enemy to bring destruction on Israel.[17]

The evidence supports accepting 4:4–13 as an authentic Amos oracle that announces the destruction of Israel.[18] Rather than a call for covenant renewal, it is a doom oracle that defends the justice of divine punishment. Israel should have no complaints. God has given it ample opportunities to repent; now is too late. Prepare for judgment! Here an original judgment oracle was reinterpreted to a later generation as a call to repentance. It was thus made relevant for another age by redefining its terms.

Amos 5:4–5, 6, 14–15

Whether or not Amos called Israel to repent in order to avoid destruction depends largely on how these five verses are interpreted. No solution to the problem is possible without coming to grips with these texts. To clarify matters, I will state my conclusion at the outset. I am persuaded that these verses are genuine Amos oracles calling Israel to redemption through rejecting its northern sanctuaries and through seeking God in Jerusalem. I accept the majority opinion, as A. V. Hunter uses this phrase. Thomas Raitt and Hunter both wrote doctoral dissertations devoted in part to these verses.[19] My conclusions depend on but depart from their research. We shall examine carefully their arguments.

Amos 5:1–17 forms a unit of varied oracles; there is no agreement on which or how many should be attributed to Amos. Hunter argues convincingly that, after the elimination of the hymnic affirmation (vv. 8–9) and the wisdom gloss (v. 13), the present order is chiasmic.[20] The unit is framed with two exhortations (vv. 1–3 and 16–17), which are in turn followed and preceded by two exhortations (vv. 4–6 and 14–15). The central point of the unit is a woe oracle (vv. 7, 10–12). Various theories attempt to explain the placement of the hymnic affirmation. H. W. Wolff contends that it is an anti-Bethel passage to be coupled with v. 6.[21] W. Rudolph argues that the catchword "turn" led a redactor to insert vv. 8–9 after v. 7.[22]

The main issue facing the interpreter of these verses is, given their meaning within their present literary matrix, how to reconstruct as accurately as possible their original context. No doubt they were interpreted by later generations of readers as a call to repentance. Was that their original purpose?

In an article published in 1971, Raitt argues that 5:4–5, 6–7, 14–15 form an independent speech unit that he entitles a "Summons to Repentance."[23] True, in many cases in which repentance is mentioned, the past failure of the people to repent has led the prophet to proclaim a doom oracle. Raitt gives twenty-four examples of this use of repentance from Amos, Isaiah, Jeremiah, and Ezekiel. In addition to 4:4–13, which we have just examined, he cites a familiar passage from Hosea:

> Their deeds do not permit them to return
>   to their God.
> For the spirit of harlotry is within them,
>   and they know not the LORD.
>
>   (5:4)

Raitt maintains that this failure to repent, when part of a doom oracle, points to an earlier time in the prophet's ministry when the fate of the people hung in the balance and they were called to repent to no avail. "The people can be judged for a failure to repent only if they are earlier clearly called to repentance."[24] The failure of the people to repent has destined them to the dark message of total destruction so characteristic of preexilic prophecy.

The failure to repent also occasioned the oracle of salvation that demanded no response from the people. In some salvation oracles, redemption is offered with no reference to repentance. For example, in striking contrast to the passage just quoted from Hosea, is another passage from Hosea in which unconditional redemption is promised after destruction:

> How can I give you up, O Ephraim!
>   How can I hand you over, O Israel!
> How can I make you like Admah!
>   How can I treat you like Zeboiim!
> My heart recoils within me,
>   my compassion grows warm and tender.
> I will not execute my fierce anger,
>   I will not again destroy Ephraim;

> for I am God and not man,
>   the Holy One in your midst,
>   and I will not come to destroy.
>   (11:8–9)

Thus, salvation is not based on repentance.

In other salvation oracles, redemption becomes the stimulus for repentance. The people return to God because they already have been redeemed, not in order to be redeemed. From II Isaiah, we read:

> I have swept away your transgressions
>   like a cloud,
>   and your sins like mist;
>   return to me, for I have redeemed you.
>   (44:22)

Raitt argues that in the Book of Amos there are three authentic summons to repentance that originated during an early period in Amos' ministry. Amos 5:4–5 provides a perfect example of a repentance summons in which the four main elements appear in proper order:

1. *Admonition*:   (v. 4)   Seek me
2. *Promise*;   and live;
3. *Accusation*:   but do not seek Bethel
                              and do not enter into Gilgal
4. *Threat*:   for Gilgal shall surely go into exile,
                              and Bethel shall come to nought.

The oracle is prefaced by a messenger formula: "For thus says the LORD to the house of Israel." The reference to "crossing over to Beersheba" Raitt considers an intrusion into the text.

This summons to repentance has all the characteristics of a speech form. It is clearly organized, concise, vivid, and memorable. It calls Israel to preserve its life by looking to God. The exile of the cultic personnel of Bethel and Gilgal foreshadows the disaster that awaits the nation if it does not seek God.

According to Raitt, 5:6–7 is another example of a summons to repentance. It is not as clearly organized as 5:4–5:

1. *Admonition*:	(v. 6)	Seek the LORD
2. *Promise*:		and live,
3. *Threat*:		lest he break out like fire in the house of Joseph, and it devour, with none to quench it for Bethel,
4. *Accusation:*	(v. 7)	O you who turn justice to wormwood, and cast down righteousness to the earth!

This repentance summons lacks the balance and conciseness of a speech form. The reversing of the accusation and threat detracts from its effectiveness. And, if Wolff is correct, vv. 7 and 10 form a unit; v. 6, with its admonition, promise, and threat, stands alone.[25] Nevertheless, Raitt clearly judges it a call to repentance.

The form of 5:14–15 is even more disjointed:

1. *Admonition*:	(v. 14)	Seek good,
2. *Accusation*:		and not evil,
3. *Promise*:		that you may live, and so the LORD, the God of hosts, will be with you, as you have said.
4. *Second Admonition*:	(v. 15)	Hate evil, and love good, and establish justice in the gate;
5. *Second Promise*:		it may be that the LORD, the God of hosts, will be gracious to the remnant of Joseph.

Again, this summons to repentance lacks the balance, order, and conciseness of an effective speech form. Its failure to include a threat weakens its effect for securing repentance. God is pictured in the second promise as rather indecisive, an atypical portrayal for Amos. On primarily stylistic and linguistic grounds, Wolff questions the authenticity of the second promise.[26] But Raitt regards it as an authentic call to repentance.

Raitt lists twenty-six additional examples of similar prophetic speech forms. To be sure, half are found in Jeremiah. Although some scholars hold that this is due to the work of the deuteronomistic editors, the studies of both Wolff and W. L. Holladay[27] suggest that the deuteronomists betray a dependence on Jeremiah. Raitt believes it is possible that the deuteronomists base their view on valid historical traditions that God's servants, the prophets, called Israel to turn again and be saved.

The genius of the summons-to-repentance speech form, according to Raitt, is its statement of alternatives. On the one hand, if the people repent, they will be saved (the promise of redemption); on the other hand, if the people refuse to repent, they will be destroyed (the threat of destruction). The summons to repentance makes the future conditional. When the people refused to repent, the prophetic proclamation that was conditional (either doom or salvation) now became unconditional (inevitable destruction followed by unmerited salvation). Raitt's hypothesis is that the summons to repentance comes from an early period in Amos' ministry before the people rejected Amos' call "to seek God and live."

Raitt's position has much to commend it. The arguments against the authenticity of vv. 6 and 14–15 are not convincing. On the other hand, Wolff demonstrates that v. 7 forms a unit with v. 10; that unit is not part of the summons to repentance. Although the second and third examples of the speech form lack the clear organization of the first, they all call Israel to repent and avoid destruction. Still unresolved is the issue of whether these are authentic Amos oracles or the product of a redactor writing for a later age. The work of Hunter furthers the argument.

Hunter opens his discussion of these verses by listing no fewer than seven different ways they can be interpreted.[28] Five major options are as follows:

1. The majority opinion accepts the passages as a call to repentance that, if heeded, will lead to redemption.
2. Hunter cites Raitt who contends that the verses come from an early period in Amos' ministry before the people rejected the prophet's call to turn again and be saved.
3. Some scholars believe that Amos offered a last-minute opportunity for the northern kingdom to repent and be saved.[29] They acknowledge, however, that the vast majority of Amos' oracles proclaim doom and destruction. Given Israel's past failures, repentance is not likely to occur.
4. Others hold that while destruction is inevitable, these verses still point to a future period when a new community will be formed.[30]
5. Hunter's own position is that the original intention of these five

verses is to proclaim God's judgment on Israel. The unit is an ironic exhortation cast in the form of a cultic lament. His study is significant enough to warrant careful consideration.

In vv. 4–5, Amos conveyed God's word in the imperative mood: "Seek me and live." The term "to seek" (*dāraš* or *biqqēš*) is a cultic expression. The people would understand this as an instruction to visit a local sanctuary in order to consult a cultic prophet about God's will. Although one usually sought the Lord at a sanctuary to determine the divine will, here Amos referred to a cultic activity with promise of salvation. A number of passages (see Deut. 30:15, 19; Ezek. 18:5) likewise offer life to those who are faithful observers of the cultus. G. von Rad declares: "The ultimate decision between life and death was then for Israel a cultic matter, and only within the cultus did the individual receive assurance that he would have life."[31] The announcement that the cultic personnel of Gilgal and Bethel will be exiled reveals the bitter irony in the oracle. The people were exhorted to seek God at a sanctuary, but were also told that the sanctuaries will be destroyed. The exhortation thus became a doom oracle.

In v. 6, Amos declared to the people:

> Seek the LORD and live,
>   lest he break out like fire in the house
>   of Joseph,
>   and it devour *with none* to quench it
>   for Bethel.

Hunter argues convincingly that this oracle is a parody of a lament. The "lest" and "with none to" clauses are keys to the cultic lament form. Two examples make this clear:

> Mark this, then, you who forget God,
>   *lest* I rend, and *there be none* to deliver!
>   (Ps. 50:22)

> O LORD my God, in thee do I take refuge;
>   Save me from all my pursuers, and deliver me,
>   *lest* like a lion they rend me,
>   and *there be none* to rescue.
>   (Ps. 7:1–2)

Hunter claims the threatening nature of v. 6 exceeds that of vv. 4–5. The prophet exhorted the people to avoid God's destructive fire by seeking him in a sanctuary. This is the usual cultic lament. But Amos had just declared that the people are forbidden to attend the sanctuaries. How then are they to seek God? The unit reaches its climax in vv. 14–15. Once the lament is expressed (v. 6), a salvation oracle is declared. If the people truly "seek good, and not evil," "hate evil, and love good," and "establish justice in the gates," then God "may" be gracious to the "descendants"[32] of Joseph.[33] The form is undoubtedly that of a cultic salvation oracle. Yet salvation is hardly assured. The people have so violated God's ways that it is unlikely he will ever again act graciously toward them even if they now seek good, hate evil, and practice justice in the courts.

Hunter concludes that the exhortations in vv. 4–5, 6, and 14–15, taken as a unit, provide little hope for Israel. Verses 4–5 urge the people to seek God in a sanctuary, but declare the sanctuaries will be destroyed. Verse 6 is a parody of a lament; the subsequent oracle is in the salvation form (vv. 14–15), announcing a slim possibility that Yahweh will be gracious if the people establish justice at the gate. Israel has little ground for hope.

As we discussed earlier, however, most scholars spiritualize the preexilic prophets and neglect their nationalistic tendencies. We have argued that Amos condemned the northern kingdom for initiating the dissolution of the Davidic empire (Chapter 4), worshiping Yahweh at the wrong places with improper rites and calendar (Chapter 5), and founding an illegitimate monarchy that failed to establish justice in the courts (Chapter 6). In this context, these exhortations take on added meaning. What did Amos mean when he urged the northern kingdom to "Seek the Lord and live" (v. 6) and to "Seek good, and not evil" (v. 14)? Both Hunter and Mays consider 5:4–5, 6, 14–15 as irony. Amos urged the people, in cultic terms, to seek God, but at the same time declared that cultic personnel at the sanctuaries will be exiled. Hunter writes:

> The people would have understood the initial imperatives to mean that they were instructed to "seek" Yahweh through activities available at the sanctuaries. Amos denies that they can seek Yahweh in this way, even at the old pilgrimage sanctuaries of Bethel, Gilgal, and Beersheba. This is admittedly a more subtle irony than in 4:4–5, where "come to Bethel and transgress" can only be interpreted ironically.[34]

In a similar manner, Mays writes:

The sentence "Seek me that you may live" is a form of priestly *tōrā*. In the mouth of the officiating priest the exhortation was an instruction to turn to Yahweh as the source of life, to come to the sanctuary where he was present to receive the dispensation of a blessing that conferred security and prosperity.... What "Seek me" as a word of Yahweh means when the shrines are excluded is left obscure and provocative.[35]

But these verses are not ironic if Amos called Israel to seek God where he was truly to be found, at the Jerusalem temple site. Kapelrud senses the problem but avoids the solution. He contends that *dāraš* is used in two different ways. When Amos said "do not seek Bethel" (v. 5), the term is used cultically; when he said "seek me and live" (v. 4), "seek the LORD and live" (v. 6), and "seek good and not evil" (v. 14), the term refers to good morals.[36] The plea to seek God in Jerusalem would make Amos a nationalist, which Kapelrud refuses to do.[37]

The simplest and best solution lies in maintaining that Amos was consistent in his use of *dāraš*. His intention was to call the northern kingdom to seek God in Jerusalem. There the supplicant worships Yahweh appropriately and receives God's blessing. Jerusalem is the sanctuary that undergirds the Davidic monarchy, commissioned to provide God's justice in the courts. One certainly should not over-emphasize the redemptive note in the exhortations, for the declaration that the northern kingdom will be destroyed pervades the Book of Amos. Nevertheless, Amos did declare that, if the people seek God in Jerusalem and accept the justice of the legitimate Davidic court, possibly Yahweh might be gracious to the descendants of Joseph.[38] Much later, after the destruction of Israel, the conditional nature of these verses received an even greater emphasis. Future generations could read them increasingly as hopeful prophetic calls to repent.

Amos 5:21-24

Although scholars disagree concerning the preceding and succeeding verses to section 5:21–24, evidence still supports the passage as a separate unit. It is preceded by a mixed form (vv. 18–20) that opens with a woe oracle, "Woe to you who desire the Day of the LORD!" It is followed by a deuteronomistic addition (vv. 25–27). Wolff argues that v. 27 (" 'therefore I will take you into exile beyond Damascus,' says the LORD, whose name is the God of hosts.") is part of vv. 21–24, constituting the punishment for guilt.[39] On the other hand, Hunter claims vv. 21–24 is a self-contained unit; given its striking similarity to the judgment prophecies of Hosea 6:4–6 and Isaiah 1:10–17, he regards the unit as a doom oracle.[40]

Its form is that of the priestly torah. The duty of the priest was to pass judgment on the cultic purity of a sacrifice (see Lev. 22:18–. 19). But here God himself, speaking through Amos, rejects unequivocally all major cultic practices. The language of rejection is strong: "I hate . . . I despise . . . I will not savor . . . I will not accept . . . I will not look upon . . . I will not listen to." The parody of the priestly torah is delivered in the form of a negative cultic decision:

> I hate, I despise your feasts,
> I will not savor[41] your assemblies.
> For though you bring to me burnt offerings
> Your gift offerings I will not accept,[42]
> and your communion meals of fatted calves
> I will not look upon.
>
> (vv. 21–22)

This is followed by a negative admonition:

> Take away from me the noise of your songs;
> to the melody of your harps I will not listen.
>
> (v. 23)

The unit climaxes in a positive admonition:

> But let justice roll down like waters,
> and righteousness like an everflowing stream.
>
> (v.24; author's translation)

Was this oracle an instruction on how to live? Did it call for a response from the people? Did it offer hope that if the people turned away from the festivals God hated and began practicing justice and righteousness they would be saved? Was this a call to repentance?

Given the strong language and somber tone, there is little ground for hoping the northern kingdom will escape destruction. Such a message can hardly be considered clear instruction on how to prevent the disaster that awaited the nation.

Amos 7:1–6

The Book of Amos contains five references to visions: locusts (7:1–3), fire (7:4–6), plumb line (7:7–9), basket of summer fruit (8:1–2), and altar (9:1). Attempts to establish some pattern in these five visions have been, on the whole, unsuccessful. Coote arranges them by order of the seasons of the year: the locusts swarm in early spring, the summer produces an intense heat leading to drought, the fruit ripens for gathering at the end of the summer, and the people assemble at the Bethel altar to celebrate the harvest.[43] Coote includes in his sequence even the conflict with Amaziah, which is placed between the third and fourth visions. Amos appears there as a dresser of sycamore trees, a summer occupation appropriately placed between the intense summer heat and ripened summer fruit. But the plumb line defies seasonal placement.

The most obvious organizational element in the text is, in the first two visions, Amos' intercession on behalf of the northern kingdom followed by God's withdrawal of destruction; in the third and fourth visions, Amos' nonintercession for Israel is followed by God's ominous words, "I will never again pass by them." In both form and location, the fifth vision has no relation to the other four visions. The intrusion between the third and fourth visions of the conflict with Amaziah at Bethel (7:10–17) is best explained by the similarity of v. 9 ("I will rise against the house of Jeroboam with a sword") and v. 11 ("Jeroboam shall die by the sword"). Verse 9 was most likely coupled with the section containing v. 11 during the collection stage of the composition. Or perhaps the Amaziah section is a redactional reinforcement of irrevocable doom. The condemnation of king and priest makes specific Amos' two doom visions.

Ernst Würthwein, in an article published in 1950, argues that Amos began as a salvation prophet. He interceded for Israel in the first two visions to avert that nation's destruction, and he proclaimed doom to her foreign enemies (oracles against foreign nations). Later in his ministry, Amos became a prophet of judgment. He refused to intercede for Israel (the third and fourth visions), and he announced God's punishment of the rebellious people.[44] Würthwein's emphasis on intercession and two periods in Amos' life are helpful, but his distinction between a prophet of salvation and a prophet of doom is overdrawn.

The first two visions may be classified as "event visions." In the first (7:1–3), Amos saw a swarm of locusts destroying the land. The crop is identified as one grown after the king had procured the first cutting to support his military establishment; if the second crop is destroyed, the people will starve. This is a vision of potential devastation. Because Amos designates a specific crop, the prophet may actually have witnessed a locust plague that suggested to him God's coming wrath on the northern kingdom. The sequence of tenses, however, shows the visionary nature of the episode. For *after* the locusts had devoured the grass, Amos interceded and God withdrew the destruction. This could only occur within the realm of a vision. In the second vision (7:4–6), Amos saw a drought so dreadful that not only was the land left parched, but the subterranean water was also desicated. Again, this could only be a vision, because when Amos interceded for the nation, the drought did not come.

In these first two visions, Amos was an active participant. He saw the visions, immediately grasped that they represented God's judgment of the nation, and interceded on the nation's behalf. The visions took place within the Council of Yahweh. With his inner eye, Amos saw what God was about to do. His intercession stopped the visions from being completed. Otherwise Amos would have had no choice but to pronounce a doom oracle releasing God's wrath upon the nation. Mays writes of an interrupted vision:

> The event belongs completely to the realm of vision, is not yet an actual event. . . . what Yahweh prepared in heaven will inexorably unfold on earth. The timing of his appeal indicates that, were the event to be completed in the vision, its re-enactment on earth would be an accomplished fact, a decree that could not be turned back.[45]

The next two visions—the plumb line (7:7–9) and the basket of summer fruit (8:1–2)—are strikingly different. In both, Amos was a passive recipient of the message. Neither vision immediately conveys a doom message. Both must be interpreted first by Yahweh. Amos cannot intercede because he did not know what the vision meant until Yahweh told him. By then the vision has been completed: Yahweh has made his decision, and Amos has been compelled to prophesy doom and destruction. The nation cannot escape God's wrath.

The meaning of the fourth vision is based on a word play, a favorite device among the Hebrew prophets. Amos saw a "basket of summer fruit" (*qayiṣ*) and God told him "the end" (*qēṣ*) has come upon the northern kingdom. The "end" may refer to the close of the year immediately before the New Year enters. John D. W. Watts interprets its meaning as that this would occur in the late summer and thus construes 8:8–14 as a reversal of fate proclaimed at a New Year Festival.[46] In any case, God through Amos is proclaiming the end of Israel. Coote believes the plumb line (*'ănāk*) also was a word play for either myself (*'ānōkî*) or you (ungrammatical form with *'ennāk*), hence reading: "I am about to set myself (or you) in the midst of my people." But the word play dropped from the text and "plumb line" appears twice.[47] Although the theory is possible, a safer assumption is to recognize that the plumb line is simply an object that measures whether a wall is straight or crooked. A crooked wall (i.e. a crooked nation) is worthless and will be destroyed.

What is the significance of Amos' intercessions on behalf of the northern kingdom in the first two visions? The tradition on which Amos' intercessions are based harks back to the Exodus 32–34 narrative in which Moses interceded for the people of Israel to avert their destruction. The parallels are so numerous that they can hardly be accidental. The following list of parallels uses Exodus 32–34 as a guide to interpretation:

1. Exodus 32–34 is a redactional unit, based on Yahwistic and Elohistic material, which centers around Israel's apostasy related to the golden calf. A connection between this narrative and the calf at Bethel symbolizing the waywardness of the northern kingdom has long been maintained.
2. Exodus 32 states that after making the golden calf, the people offered sacrifices to Yahweh, "sat down to eat and drink, and rose to play" (v. 6), and further, that as Moses descended the mountain, he heard the sound of singing and saw the people dancing (vv. 18–19). Could this not be a reference to the *marzēaḥ* festival that Amos condemned?
3. In Exodus 32–34, because of Israel's apostasy, God decided to destroy the people.

And the LORD said to Moses: "I have seen this people; now behold, it is a stiff-necked people, now therefore let me alone, that my wrath may burn not against them and I may *consume* them. (32:9–10)

Similarly, in the Book of Amos, God's wrath upon Israel comes in the visions of the locusts that had finished *eating* the grass of the land and a devastating drought that *consumed* the great deep, *was eating up* the land. All three words are derived from the verb *'ākal*, "to consume," "to eat."

4. In both Exodus 32–34 and Amos 7:1–6, Yahweh planned to destroy his people, but left open a hope for intercession. Both Moses and Amos stepped into the breach and prevented the nation's annihilation.

5. Moses ascended the theophanic mountain of Sinai, near the throne of God, to intercede for the people; Amos interceded by entering the Council of Yahweh.

6. In Exodus 32–34, Moses interceded by reminding God of his promises to Abraham, Isaac, and Israel/Jacob (32:13). Not accidentally, Amos chose to call Israel by its old patriarchal name of Jacob, thereby reminding God of his past promises to his people. In addition, Brueggemann demonstrates how appropriate is the description of Jacob as "small," because here is a reminder of the patriarchal narratives in which Jacob is small (younger) and Esau is great (older) (Gen. 27:15, 42), yet Jacob receives the inheritance.[48] Both Moses and Amos reminded Yahweh of Israel's dependence on him.

7. Exodus 32:14 records God's response: "And Yahweh *repented* (waw consecutive imperfect of *niham*) concerning the evil which he said he would do to his people." The response of God to the intercession of Amos is similar: "The LORD *repented* (perfect of *niham*) concerning this: 'It shall not be,' said the LORD" (7:3, 6).

8. Moses continued to intercede throughout the unit (32:11, 30; 33:12; 34:8), therefore God *passed* before him and declared his mercy upon the people (34:6). Amos did not intercede in visions three and four, therefore God declared he would never again *pass* by them (7:8; 8:2). The destruction of the northern kingdom is inevitable. Both texts use the verb *'ābar*, "to pass over." Amos employed once again a reversal motif.

9. In both Exodus 32–34 and Amos 7:1–6, Yahweh changed his mind and forgave Israel on the basis of Moses' and Amos' intercessions (Exod. 32:14; 33:19; 34:6; Amos 7:3, 6). This is not an example

of cheap grace. The Exodus narrative concludes with a new revealing of the law by which the people should order their lives. Nor does Amos' intercession for Israel ask for forgiveness without responsibility. The now-conditioned reprieve demands that Israel practice justice and righteousness.

Did Amos call Israel to repent in order to escape destruction? The evidence is not clear. Certainly the deuteronomists and the deuteronomistic historians believed destruction came upon Israel and Judah because the people failed to heed the prophetic call for repentance. There is little within Amos on which to base such a call. Amos 4:4–13, often quoted as a call to turn about and be saved, is actually a doom oracle. The same is true of the famous call for justice and righteousness in 5:21–24. The language is too negative to find comfort there.

This leaves 5:4–5, 6, 14–15, and 7:1–6. Do these verses provide any basis for believing Amos called the northern kingdom to turn about and be saved? I believe they do.[49] These verses point to an early period in Amos' ministry when the prophet interceded for the northern kingdom and urged God to withhold his destructive wrath. Amos believed that if the northern kingdom would only turn from its corrupt sanctuaries, worship Yahweh in Jerusalem, acknowledge the legitimate Davidic ruler, and live under the justice of the royal courts, then all would go well for them.

It was not to be. Amaziah's rejection of Amos at Bethel epitomizes the prophet's reception in the north. With his summons to repentance rejected, Amos had no alternative save to proclaim a dark message of Israel's destruction. Did he see any hope for the northern kingdom after its defeat and exile? Did Amos then proclaim only that nation's complete annihilation? Or did he accept a belief in a remnant that would return?

8

The Destiny of the Northern Kingdom of Israel

DID AMOS BELIEVE that Israel had forfeited its right to be God's people? Had its failure to repent caused God to issue a death sentence on that nation? Or did Amos see some hope beyond destruction when God's election of Israel would still be valid? The answer to these questions depends largely on how one interprets the concept of the Day of Yahweh in Amos, and whether any verses in 9:8b–15 can be attributed to Amos.

A majority of scholars consider Amos a prophet of doom proclaiming the total destruction of Israel. They argue that Amos saw no future for the northern nation except devastation and annihilation. For them, the Day of Yahweh in Amos is a day of total darkness with no glimmer of light. These scholars believe that the final, authentic words of Amos are:

> Behold, the eyes of the Lord God
> are upon the sinful kingdom,
> and I will destroy it from the surface
> of the ground.
>
> (9:8a)

What then follows (9:8b–15) becomes a postexilic hope addition for encouraging those in the exile and those in the struggling, restored community to await a better day.

To contrast Amos with Hosea at this point is not uncommon. Hosea is a prophet of hope who believed God would restore Israel after its punishment by destruction and exile. For Hosea the Day of Yahweh had a double meaning: the destruction of the nation followed by restoration to the land.

We have argued, however, that Amos foresaw that the righteous poor of the northern nation would experience redemption in a re-united nation under a just Davidic ruler. This chapter explores the evidence supporting that thesis.

Three Theories on the Day of Yahweh

The concept of the Day of Yahweh has long been recognized as central to the prophetic view of the future.[1] The Book of Amos contains the oldest prophetic reference to it:

> Woe to you who desire the day of the LORD
>> Why would you have the day of the LORD?
> It is darkness, and not light;
>> as if a man fled from a lion,
>>> and a bear met him;
>> or went into the house and leaned with his
>>> hand against the wall,
>> and a serpent bit him.
> Is not the day of the LORD darkness,
>> and not light,
>> and gloom with no brightness in it?
>
>   (5:18–20)

The language of the passage clearly acknowledges that certain factions of the people looked forward to the Day of Yahweh. The period was popularly associated with light and brightness, with a time when God would bless his people. But widespread disagreement prevails concerning the precise meaning of the Day of Yahweh. In recent years, three theories have dominated the field.

Mowinckel and the Mythical Approach

The Scandinavian scholar, Sigmund Mowinckel, maintains that the Day of Yahweh must be interpreted within the context of Israel's cult. The evidence for this view Mowinckel finds in the Hebrew psalms, the hymnbook for temple worship. Focusing on the cult in the monarchical period, he finds the Day of Yahweh central to the

annual New Year Festival when the people celebrated the enthronement of Yahweh in myth and ritual.

> The LORD reigns; he is robed in majesty;
>   the LORD is robed, he is girded with strength.
> Yea, the world is established;
>   it shall never be moved;
> thy throne is established from of old;
>   thou art from everlasting.
>
> The floods have lifted up, O LORD,
>   the floods have lifted up their voice,
>   the floods lift up their roaring.
> Mightier than the thunders of many waters,
>   mightier than the waves of the sea,
>   the LORD on high is mighty!
>
> Thy decrees are very sure;
>   holiness befits thy house,
> O LORD, for evermore.
>
> (Ps. 93; see also Pss. 47, 95–99)

In language reminiscent of Marduk's triumph over Tiamat and Baal's
defeat of Yamm, Yahweh overcomes chaos and establishes justice,
peace, and prosperity for his people. In history, God gives victory
over enemy forces; in nature, he sends the rains essential for abundant
harvests. The agent through whom God acts is the king; as God's
representative, the king ensures the well-being of the nation by ruling
with justice and righteousness. On the annual Day of Yahweh, peace
and prosperity are reaffirmed for successive generations.

Mowinckel summarizes his position:

> The whole picture of the future can therefore also be summed up in
> the expression, *the day of Yahweh*. Its original meaning is really the
> day of His manifestation or epiphany, the day of His festival, and
> particularly that festal day which was also the day of His enthrone-
> ment, His royal day, *the* festival of Yahweh, the day when as king He
> came and "wrought salvation for his people." As the people hoped
> for the realization of the ideal kingship, particularly when reality fell
> furthest short of it, so, from a quite early period, whenever they were
> in distress and oppressed by misfortune, they hoped for and expected
> a glorious "day of Yahweh" (cf. Amos v, 18 ff.), when Yahweh must

remember His covenant, and appear as the mighty king and deliverer, bringing a "day" upon His own and His people's enemies (cf. Isa. ii, 12 ff.), condemning them to destruction, and "acquitting" and "executing justice" for His own people.[2]

Von Rad and the Historical Approach

The German scholar Gerhard von Rad understands the Day of Yahweh within a different context; he views the term within the historical traditions of holy war, especially those wars fought to conquer the land. Von Rad focuses his study on the cultic practices of the premonarchical period when, at major shrines like Shechem and Shiloh, the people celebrated Yahweh's victory over the Egyptian pharaoh and his conquest of Canaan. There have been and will be many Days of Yahweh, many times when God gives his people victory over their enemies. The Day of Yahweh includes a summons to battle, the marching to war of armies, and Yahweh's personal entrance onto the battlefield. The Day of Yahweh is often accompanied by such cosmic phenomena as earthquake, thunder, hail, and eclipse.

For example, a poem on Babylon in Isaiah 13 begins with the summons to war (vv. 2–4); then come the fighting men from afar to battle Babylon (v. 5). Next a cosmic upheaval occurs.

> Behold, the day of the LORD comes,
>   cruel, with wrath and fierce anger,
> to make the earth a desolation
>   and to destroy its sinners from it.
> For the stars of the heavens and their
>   constellations will not give their light;
> the sun will be dark at its rising
>   and the moon will not shed its light.
>   (vv. 9–10)

God himself enters the battle and the scene depicts carnage.

> Whoever is found will be thrust through,
>   and whoever is caught will fall by the sword.
> Their infants will be dashed in pieces
>   before their eyes;

> their houses will be plundered and their
> wives ravished.
>
> (vv. 15–16)

Yahweh will punish Babylon for her ruthless oppressions of other people.

The following quotation from von Rad highlights the conflict between Mowinckel's and von Rad's positions:

> [T]he proper procedure for investigation is first to rule out all possible interpretations based on a rather far-fetched mythology, and then to ask whether Israel herself did not have, in her own old traditions, some knowledge of the concept of Jahweh's coming specifically to wage a war, with its accompaniment of miraculous phenomena. This is, of course, the case. In itself, the almost stereotyped connexion of the day of Jahweh with intervention in war reminds one of the holy wars and all the phenomena which traditionally accompanied them. In this concept of Jahweh's coming to an act of war we have at least one concept clearly stamped with Israel's own tradition, and we should establish its relationship with the prophetic utterances about the day of Jahweh before we try any other methods of interpretation.[3]

Mowinckel's theory stresses the myth of God's victory over chaos expressed through a yearly recreation of the enthronement of Yahweh within the cult. This cultic reinforcement by the people resulted in their symbolic recreation and reconfirmation of the cosmos itself. Von Rad's theory, on the other hand, emphasizes the premonarchical holy wars that were remembered and rehearsed at the various confederation shrines.

Cross and the Mythico-Historical Approach

The American scholar Frank Cross argues that both the historical theme of conquest through holy war and the mythological elements of Yahweh's enthronement found their places in cultic life during the monarchy. Cross's theory is especially evident in his study of the divine warrior motif. Yahweh as divine warrior fights on the historical plain to overcome Israel's enemies; Yahweh as the enthroned king, described in divine warrior imagery, fights on the cosmic plain to defeat the forces of chaos. Israel's festival life during the monarchical period celebrated God's victories in both realms. The two realms blend when historical events become mythologized or when mythology becomes historicized. A striking example appears in Isaiah 51:9–10:

> Awake, awake, put on strength,
> O arm of the LORD;
> awake, as in days of old,
> the generations of long ago.
> Was it not thou that didst cut Rahab in pieces,
> that didst pierce the dragon?
> Was it not thou that didst dry up the sea,
> the water of the great deep;
> that didst make the depths of the sea
> a way for the redeemed to pass over?

Here the prophet began by reflecting on the creation myth of God piercing the sea dragon Rahab or Yamm. The sea references reminded him of the historical redemption of the Hebrew people at the crossing of the sea. Thus mythological creation and historical deliverance marvelously blended to reveal God's mighty powers. Just as God split the dragon to create order from chaos, so he split the seas to create a nation of his people.[4]

For Cross, the Day of Yahweh is both historical and mythological. The holy-war imagery of the conquest (von Rad) was joined with the enthronement of Yahweh in the royal cult (Mowinckel). He writes:

> The Day of Yahweh is the day of victory in holy warfare; it is also the Day of Yahweh's festival when the ritual Conquest was enacted in the procession of the Ark, the procession of the King of Glory to the temple, when "God went up with the festal blast, Yahweh with the sound of the horn . . . for Yahweh is king of the whole earth."[5]

Cross' mythico-historical approach to the Day of Yahweh through the divine warrior motif is the theory most helpful in analyzing Amos' view of the Day of Yahweh.

The Day of Yahweh in Amos

Amos' proclamation that the Day of Yahweh was a day of darkness and gloom, not light and brightness (Amos 5:18–20), must have shocked his audience. The people longed for the coming of the Day of Yahweh, because they believed it promised peace and prosperity. Employing dramatic reversal in the form of a simile, Amos likened the Day of Yahweh to a time when people flee from a lion only to encounter a bear. If they attempt to escape danger by entering a house, a snake strikes them. The meaning is clear: to escape destruction is impossible, because catastrophe awaits.

In addition to the title "Day of Yahweh" (*yôm-YHWH*), which occurs three times in Amos 5:18–20, four other terms also refer to that same day. "In that day" (*bayyôm hahû*) appears five times (2:16; 8:3, 9, 13; 9:11),[6] whereas "on the day" (*běyôm*: 3:14), and "evil day" (*yôm rā*: 6:3), and "the days are coming" (*yamim bā'îm*: 4:2)[7] are each used once. There are also numerous passages which, though not containing these technical terms for Day of Yahweh, still employ imagery related to that day. In depicting this tragic day of doom, Amos used considerable literary skill. We shall examine the five characteristics of the Day of Yahweh in Amos: fire, exile, natural catastrophes, lamentations, and total destruction.

Fire

Eight times Amos referred to a "fire" that "devours" God's enemies. The expression is used in each of the oracles against foreign nations (1:4, 7, 10, 12, 14; 2:1), as well as against Israel (5:6; 7:4). What did Amos mean? Possibly Amos was depicting the natural phenomenon of fire used in the burning of enemy strongholds. With this interpretation, 7:4 refers to a consuming midsummer drought that dries up the subterranean waters and causes devastation in the land. The references would be to natural catastrophes, specifically fire in warfare and a severe drought.

More likely, however, the "fire which devours" is holy-war imagery in a divine warrior context. In 7:4 Amos intended to convey not a natural phenomenon, but a divine fire (lightning) with the power to destroy both the subterranean deep (*těhôm*) and the land.

Thus the Lord God showed me: behold the Lord God was calling for a judgment by fire, and it devoured the great deep and was eating up the land.

This is the same "deep" as in Genesis 1:2, which God subdued in order to fashion creation (cf. Marduk vs. Tiamat and Baal vs. Yamm).

"The text is thus to be understood as a prophetic vision of the conflict of Yahweh with the primordial monster of the deep, in which his weapon is lightning, or supernatural fire."[8]

The use of supernatural fire is a characteristic tool of the warrior gods of the Ancient Near East for achieving victory over chaos. Images of Baal show a male deity, thunderbolt in hand and ready for hurling at Prince Yamm.[9] Yahweh, Israel's divine warrior, is similarly portrayed in all his majesty by Habakkuk 3:2–16; especially relevant for an analysis of Amos' use of fire are the following verses:

> His brightness was like the light,
>   rays flashed from his hand;
>   and there he veiled his power.
>   (v. 4)

> Was thy wrath against the rivers, O LORD?
>   Was thy anger against the rivers,
>   or thy indignation against the sea,
> when thou didst ride upon thy horses,
>   upon thy chariot of victory?
>   (v. 8)

> The mountains saw thee, and writhed;
>   the raging waters swept on;
> the deep gave forth its voice,
>   it lifted its hands on high.
>   (v. 10)

> Thou didst pierce with thy shafts the head
>   of his warriors,
>   who came like a whirlwind to scatter me,
>   rejoicing as if to devour the poor in secret,
> Thou didst trample the sea with thy horses,
>   the surging of mighty waters.
>   (vv. 14–15)

The manifestation of God in Psalm 97 is also similar:

> The LORD reigns; let the earth rejoice;
>   let the many coastlands be glad!
> Clouds and thick darkness are round about him;
>   righteousness and justice are the
>   foundation of his throne.
> Fire goes before him,
>   and burns up his adversaries round about.

His lightnings lighten the world;
the earth sees and trembles.
The mountains melt like wax before the LORD,
before the LORD of all the earth.
(vv. 1–5)

In two of Amos' oracles against foreign nations, the divine-fire imagery combines with holy-war language.[10] Amos 1:14 adds "with shouts in the day of battle, with a tempest in the day of the whirlwind" and 2:2 includes "and Moab shall die amid uproar, amid shouting and the sound of the trumpets."

In the past the Hebrew people had been blessed when their God had used his divine powers to overcome their enemies and to give them their promised land. They now took pleasure in Amos' proclamation of destruction on the surrounding hostile foreign nations. How the people of Israel must have raged when Amos warned that they had become God's enemy; the consuming fire that once assured deliverance would now destroy them.

Exile

A second characteristic of the Day of Yahweh pertains to Amos' unwavering contention that Israel would be destroyed and that exile would follow for those people who survived. Amos did not designate the nation that God would employ to destroy the northern strongholds and cities. He simply referred to the destroyer as "an adversary" (3:11) and "a nation" (6:14). There was no doubt that the devastation would be overwhelming. Even foreign nations would gather to witness the tragedy (3:9) and testify against the northern kingdom for the crimes it had committed (3:13). The description of warfare in the Book of Amos is graphic.

An adversary shall surround the land,
and bring down your defenses from you,
and your strongholds shall be plundered.
(3:11)

The high places of Isaac shall be
made desolate,
and the sanctuaries of Israel shall
be laid waste,

> and I will rise against the house of
> Jeroboam with the sword.
>
> (7:9)

The cities of the northern kingdom will send forth troops to fight the enemy, but only a handful will survive the battle.

> The city that went forth a thousand
>   shall have a hundred left,
> and that which went forth a hundred
>   shall have ten left
>   to the house of Israel.[11]
>
> (5:3)

But the most tragic aspect of the Day of Yahweh is not the destruction, but exile from the land.[12] Yahweh had chosen them to be his people (3:2) and had destroyed their enemies so that they might possess the land (2:9). Now they will be treated as God's enemy; they will lose their inheritance and become landless exiles in a foreign nation.

> Jeroboam shall die by the sword,
>   and Israel must go into exile
>   away from his land.
>
> (7:11)

Amos' condemnation of Amaziah promised what every person dreads as a consequence of war, specifically family tragedies, loss of home, and deportation.

> Your wife shall be a harlot in the city
>   and your sons and your daughters shall
>   fall by the sword,
>   and your land shall be parceled out by line;
> you yourself shall die in an unclean land,
>   and Israel shall surely go into exile
>   away from its land.
>
> (7:17)

Those deported should take no comfort in being spared their lives; in exile the enemy will slay them (9:4).[13]

Natural Catastrophes

Amos further proclaimed that on the Day of Yahweh the northern kingdom would experience God's wrath also through the forces of nature. As the Book of Amos attests, the prophet's words came two years before a major earthquake (1:1). Any person who has experienced such a natural catastrophe knows the event is filled with dread. Because Amos' oracles against the northern kingdom contain an inordinate number of references to a natural catastrophic phenomenon, a major earthquake may well have seemed to Amos an appropriate sign of God's displeasure with his people.

> I will smite the winter house with the
>   summer house;
>   and the houses of ivory shall perish,
> and the great houses shall come to an end.[14]
>
>   (3:15)

In a curious use of mixed imagery, Amos compared the shifting of land during an earthquake to the rise and fall of the Nile during its annual inundation (8:8). In another carefully crafted oracle, he declared that the quaking earth will terrorize even mighty warriors and that those bowmen who ride in chariots behind their drivers will be unable to keep their feet.

> Flight shall perish from the swift,
>   and the strong shall not retain his strength,
>   nor shall the mighty save his life;
> he who handles the bow shall not stand,
>   and he who is swift of foot shall not
>   save himself,
>   nor shall he who rides the horse save his life,
> and he who is stout of heart among the mighty
>   shall flee away naked in that day.
>
>   (2:14–16)

Amos mentioned other natural catastrophes, but did not develop any as fully as the earthquake. In the visions, God decided to withhold a locust plague (7:1–3) and a destructive drought (7:4–6). Yet on another occasion, Amos spoke of a drought coming with such intensity that even strong youths would faint from lack of water (8:13).

One oracle even predicts an eclipse (8:9) that, while causing no disaster, is capable of producing terror.

Amos 6:9–10 is a curious passage best interpreted as a plague upon the land.[15] Death is widespread; corpses fill the house. A relative comes to remove the bodies. Another person has gone into an inner room looking for more corpses. When someone is found alive, he is told not to call to God for help because Yahweh's presence will only lead to further disaster. The tone of the passage is sufficiently dreadful that readers are reminded of descriptions of the Black Plague in medieval Europe or of the Warsaw Ghetto and Auschwitz in more recent times. What a frightening way to convey the darkness and gloom of the Day of Yahweh! Dante's inferno looms no worse.

Lamentation

Such widespread devastation by warfare and natural disasters calls for great lamentation, a further characteristic of the Day of Yahweh. Amos found the funeral dirge an appropriate expression for the somber nature of the time.

Amos 5:1–2 and 5:16–17 form a funeral lament framing a series of Amos' oracles. Although God had passed over his people in the plagues of the Exodus event and punished only the Egyptian pharaoh and his people, now God will pass through the northern kingdom bringing destruction on farmlands and vineyards. Amos no longer intercedes for them; the doom visions of plumb line (7:7–9) and basket of summer fruit (8:1–2) will come to pass. There will be great wailing in that day; corpses will be too numerous to bury (8:3). Although Amos 8:10 is a description of mourning rites, most scholars regard this verse as redactional.

Total Destruction

Who will go into exile from Israel? Given Amos' message of social justice, the poor and the needy would probably escape destruction and inherit the land.[16] Numerous passages indicate that on the Day of Yahweh divine wrath will be visited upon the rich. Only the wealthy have the means to own chariots, hire riders, and fight with bows (2:13–16). Moreover, those who inhabit fortresses and "store up" violence are the upper classes (3:9–11). Only people in the upper classes owned two homes, one for winter and one for summer (3:15a).

They only possessed houses built with expensive hewn stone (5:11) and houses with ivory inlay furnishings (3:15b). Certainly only the upper classes have the wealth to bribe the courts (2:6–7; 5:12) and to engage in other oppressive economic practices against the poor (2:6–8; 4:1; 5:7, 10–11, 15, 24; 8:4–6). Only that class participates in the *marzēaḥ* festival (4:1, 6:4–7).

Yet numerous passages in Amos also declare the total destruction of the northern nation, the poor as well as the wealthy. Small comfort when Amos compared those who survive in Samaria with the two legs and piece of an ear a shepherd retrieves from a sheep killed by a lion (3:12). For Amos declared elsewhere that the entire city of Samaria and "all that is in it" would be destroyed (6:8). Not only will the great houses be torn down, but also the small ones belonging to the poor of the land (6:11). Disaster will pervade the entire land (6:14b); none shall escape judgment (9:1). The classic reference to the Day of Yahweh in 5:18–20 forebodes total darkness, no brightness for anyone. Many scholars postulate that the authentic words of Amos end with his declaration that Yahweh will destroy Israel from the surface of the ground (9:8a).

Amos and the Reconstituted Davidic Empire

The central argument of this book has been that insufficient attention has been given to the relationship between nationalism and religion in the studies of the history and religion of Ancient Israel. This is especially true in content analyses of the preexilic Hebrew prophets. Nationalism has been viewed simply as one element of postexilic particularism (see Ezra and Nehemiah), detracting from Judaism's rich heritage. Even if national religion existed in the monarchical period, most scholars argue that the Hebrew prophets stood prominently above their contemporaries, declaring that Yahweh is the God of all nations, calling all peoples to practice social justice and righteousness. Most scholars claim that preexilic Hebrew prophecy sought to free the Hebrew people from the narrow confines of nationalism by proclaiming a universal faith for all humankind.

We have shown in our examination of Amos, however, that religion in the preexilic period was intimately related to the nation-state. State religion in Israel began with the establishment of the Davidic dynasty when Jerusalem was made the religious and administrative center of the kingdom. That city with its holy mount, sacred temple, official priests, and festal calendar supported those religious beliefs and practices receiving governmental approval. Jerusalem was also the geographic site for the administrative, central offices of the state, a state that had gradually become, under David's leadership, an empire with special relationships among its immediate neighbors. Israel's state religion was focused in the Davidic monarchy; both cultic officials and government bureaucrats acknowledged the king as their official leader.

That Amos accepted this form of state religion is not surprising, for he had already experienced a similar pattern in his native land of Judah. For him Jerusalem was the main cultic center for the official religion. His own city of Tekoa was administered by officials responsible to the king who ruled from Jerusalem. This was the proper way to run a kingdom. What Amos opposed was the crumbling of that empire through the rebellion of the foreign nations having personal relationships with David; he especially condemned the rebellion of the northern kingdom that established a nation to rival Judah. Having rejected both God's choice of the Davidic line of kings and the uniqueness of Jerusalem as a worship site, the northern rebels replaced the legitimate Davidic rulers with kings lacking legitimacy, established their own capital of Samaria, and instituted state religion at old Canaanite shrines such as Bethel. The resulting judicial system administered by the government at Samaria was corrupt and oppressive to the poor and the needy.

Amos called the northern kingdom to repent, to reunite with the south, to accept once again the just rule of the Davidic kings, and to practice the true form of religion embodied in the Jerusalem cult. Only in this way, he said, would the north escape destruction. When the people rejected his message, he had no alternative but to proclaim the total destruction of the northern kingdom.

Yet Amos still asserted hope for those taken into exile. Chapter 4 presents evidence that 9:8b–12 is authentic Amos. In that passage Amos expressed his hope that after the destruction of the rebellious foreign nations that were once members of the Davidic empire, and after the defeat and exile of Israel, God would "in that day" reconstitute the Davidic empire. Here the Day of Yahweh is not a prediction of doom, but an oracle of salvation. Amos described that day as a time of return to the land by the righteous, the true "house of Jacob," whom he had identified with the poor and the needy. In that reborn nation, true religion centered in Jerusalem would be honored, and the ideal Davidic king would establish a proper judicial system to ensure that justice roll on like waters and righteousness as a never-ceasing stream.

Notes

Chapter 1

1. Though undeveloped, the basic thesis of this book is found in G. Henton Davies, "Amos—the Prophet of Re-Union." I am grateful for the encouragement Dr. Davies gave me in my research on Amos. Arvid S. Kapelrud, in "New Ideas in Amos," leans toward this approach: "Amos is here the spokesman of Yahwistic circles in the South, whose reaction against what was going on in leading layers of the northern kingdom was very strong" (p. 197). Robert Gordis, "The Composition and Structure of Amos," believes the Book of Amos consists of two collections: in the first collection Amos hoped for the national salvation of Israel, and in the second collection, written after the Amaziah conflict, he saw only the destruction of Israel. However, a reconstituted Judah will incorporate the faithful from the destroyed Israel. See also Klaus Koch, *The Prophets: Volume One, The Assyrian Period*, pp. 36–76, with which I am in substantial agreement. Compare Klaus Koch, *Amos: Untersucht mit den Methoden einer strukturalen Formgeschichte*.

2. For example, J. Philip Hyatt, *Prophetic Religion*, p. 151; Julius A. Bewer, *The Prophets*, p. 470; T. Henshaw, *The Latter Prophets*, pp. 90–92; Curt Kuhl, *The Prophets of Israel*, pp. 64–65; Emil G. Kraeling, *The Prophets*, pp. 44–45.

3. For a detailed discussion of this approach to Amos' attitude toward foreign nations, see Norman K. Gottwald, *All the Kingdoms of the Earth: Israelite Prophecy and International Relations in the Ancient Near East*, pp. 94–119.

4. See "The Pittsburgh Platform (1885)," in Nathan Glazer, *American Judaism*, pp. 151–52.

5. The modern State of Israel, however, has never clearly defined

whether it is a secular or a religious state. It wavers depending on the crisis at hand.

6. A number of books are available on methodology in the study of the Hebrew Scriptures. See Claus Westermann, *Basic Forms of Prophetic Speech*; John H. Hayes, ed., *Old Testament Form Criticism*. Fortress Press has its Guide to Biblical Scholarship Series: Gene M. Tucker, *Form Criticism of the Old Testament* (1971); Norman C. Habel, *Literary Criticism of the Old Testament* (1971); David Robertson, *The Old Testament and the Literary Critic* (1971); Walter E. Rast, *Tradition History and the Old Testament* (1972); Ralph W. Klein, *Textual Criticism of the Old Testament: The Septuagint after Qumran* (1974); Edgar Krentz, *The Historical-Critical Method* (1975); J. Maxwell Miller, *The Old Testament and the Historian* (1976); H. Darrell Lance, *The Old Testament and the Archaeologist* (1981); Robert R. Wilson, *Sociological Approaches to the Old Testament* (1984); James A. Sanders, *Canon and Community: A Guide to Canonical Criticism* (1984).

7. The verb used for "seeing" is *ḥāzâ*, "to have a vision/perceive a vision" and not the more standard verb *raʾâ*. The use of *ḥāzâ* raises the question of whether *dĕbārîm* should be translated "words" or "things, matters," thus including visions as well as auditions. In 7:12 Amaziah called Amos a "seer," employing the participle *ḥōzeh*. The editor may be establishing a conceptual connection between chapters 1–6 and 7–9. (For this suggestion, I am indebted to E. Theodore Mullen, Jr.) This and all subsequent quotations from Hebrew Scriptures are taken from the RSV unless otherwise stated.

8. Dating follows the chronology presented in John Bright, *A History of Israel*. The more inclusive abbreviations B.C.E. (Before the Common Era) and C.E. (Common Era) are used instead of the Christian-oriented abbreviations B.C. and A.D.

9. It is customary to use the term "deuteronomic" to designate the legal code and its framing passages contained in the Book of Deuteronomy and "deuteronomistic" to designate the history of Israel contained in Joshua through Kings along with its introductory framework in Deuteronomy.

10. For an informative discussion of redactionism that focuses on Amos, see Bruce Vawter, "Prophecy and the Redactional Question," pp. 127–39.

11. On the deuteronomistic redaction of Amos, see W. H. Schmidt, "Die deuteronomistische Redaktion des Amosbuches." For a critique of Schmidt's argument, see T. R. Hobbs, "Amos 3, 1b and 2, 10."

12. However, recounting the life of Jeroboam II (II Kings 14:23–29), the deuteronomistic historians did mention a prophet, Jonah (II Kings 14:25), who predicted the expansion of Israel. In keeping with their view of history, they found this prophecy fulfilled in the reign of Jeroboam II. These verses may have been written during the exilic period when Israel posed no threat to Judah. But the writers betray their basic dislike of Israel by allowing

Jeroboam II only seven verses, indicating that he "did what was evil in the eyes of the LORD" and, if the corrupt Masoretic text is followed, that the empire he ruled once belonged to Judah. Frank Crusemann, "Kritik an Amos im deuteronomistischen Geschichtswerk: Erwägungen zu 2. Könige 14:27," argues that the theological position of the deuteronomistic historians prevent them from accepting the radical rejection of Israel by Amos as the word of Yahweh.

13. Gottwald, *All the Kingdoms of the Earth*, pp. 97–100. Robert B. Coote, in *Amos among the Prophets: Composition and Theology*, pp. 19–24, believes the A stage of Amos should be dated during the reign of Tiglath-pileser III, ca. 738. The main reasons given for this late date are:

1. Jeroboam II is not mentioned in any of the authentic oracles as identified by Coote.
2. The social history implied by the authentic oracles is not a period of peace but a period of turmoil; in other words, the period of Tiglath-pileser's initial campaigns in the west when the deportation of entire populations was introduced.

14. Hans Walter Wolff, *Joel and Amos*, p. 151, n. 102.

15. Yigael Yadin et al., in *Hazor II: An Account of the Second Season of Excavation*, pp. 24–26, 36–37, dated the earthquake ca. 760, but later revises it to the first half of the eighth century. See Joseph Blenkinsopp, *A History of Prophecy in Israel*, pp. 93, 127–28, n. 31.

16. Koch, in *The Prophets: Volume One, The Assyrian Period*, p. 70, suggests a Galilean Tekoa that is attested in postbiblical times.

17. Arvid S. Kapelrud, *Central Ideas in Amos*, pp. 6–7; Erling Hammershaimb, *The Book of Amos, A Commentary*, p. 17.

18. On the influence of cultic terminology on Amos, see James L. Crenshaw, "The Influence of the Wise upon Amos."

19. Peter Craigie, "Amos the *nōqēd* in the Light of Ugarit."

20. The literature examining the relationship between vision and ecstasy is extensive. See W. Jacobi, *Die Ekstase der alttestamentlichen Propheten*; Sigmund Mowinckel, "The Spirit and the Word in the Pre-exilic Reforming Prophets"; A. Guillaume, *Prophecy and Divination among the Hebrews and Other Semites*; A. Haldar, *Associations of Cult Prophets among the Ancient Semites*; H. Knight, *The Hebrew Prophetic Consciousness*; W. C. Klein, *The Psychological Pattern of Old Testament Prophecy*; J. Lindblom, *Prophecy in Ancient Israel*. The use of the term "spirit" in Hebrew prophecy is instructive. Spirit is associated with both preclassical ecstatic prophets and postexilic prophets, but rarely with the preexilic classical prophets. The latter group received the word. This decline in the use of the term spirit among the preexilic classical prophets probably indicates their aversion to the excessive display of emotions by those ecstatic prophets who preceded them. For the

preexilic classical prophets, the religious experience came in the form of visions rather than ecstatic seizures. Still, they placed more emphasis on the message to be delivered than on visions. They stressed oracle over experience. However, the passage of time allowed the term spirit once again to regain popularity in the postexilic period, but without ecstatic overtones.

21. For references to the Council of Yahweh, see Genesis 1:26–27; 3:22; 6:1–4; 11:7; Deuteronomy 32:8–9; 33:2–3; I Kings 22:19–23; Job 1–2; Psalms 29; 58:1; 82; 89; Isaiah 6; 24:21; 40:1–11; Jeremiah 23:18–22; Ezekiel 1:1–3:15; Zechariah 3; Daniel 10. On the place of the council in Hebrew thought, see H. Wheeler Robinson, "The Council of Yahweh"; R. N. Whybray, *The Heavenly Counsellor in Isaiah xl 13–14: A Study of the Sources of the Theology of Deutero–Isaiah*, pp. 34–38, 64–77; G. Cook, "The Sons of (the) God(s)"; Frank M. Cross, Jr., "The Council of Yahweh in Second Isaiah"; Max E. Polley, "The Call and Commission of the Hebrew Prophets in the Council of Yahweh, Examined in its Ancient Near Eastern Setting."

22. Frank M. Cross, Jr. and David N. Freedman, "The Blessing of Moses," p. 201, n. 19. For the assembly of the gods in Canaanite thought, see William F. Albright, *Yahweh and the Gods of Canaan: A Historical Analysis of Two Contrasting Faiths*, pp. 104–9; Frank M. Cross, Jr., *Canaanite Myth and Hebrew Epic: Essays in the History of the Religion of Israel*, pp. 13–43, 186–90; Marvin H. Pope, *El in the Ugaritic Texts*, pp. 27–29, 47–54; E. Theodore Mullen, Jr., *The Divine Council in Canaanite and Early Hebrew Literature*.

23. James Muilenberg, "Introduction and Exegesis of Isaiah 40–66."

24. Amos 3:7 is the most direct reference to the Council of Yahweh, but it is a deuteronomistic addition to the text. See Wolff, *Joel and Amos*, pp. 181, 187–88.

25. The RSV translates the Hebrew divine name "Yahweh" by "LORD," all capital letters. When "Lord" or "lord" appears, it translates either "*adōnāy*" or "*baʿal*"; whether the term applies to the God of Israel, to some other god, or to a human overlord or husband depends on context. "God" in the RSV translates the Hebrew general term for deity, "*ĕlōhîm*," or one it its various forms.

26. John D. W. Watts, *Vision and Prophecy in Amos*, pp. 9–12. Compare H. H. Rowley, "Was Amos a Nabi?"

27. H. Wheeler Robinson, "Prophetic Symbolism."

28. The main text for distinguishing true from false prophecy is Deuteronomy 18:15–22. This text has exerted considerable influence on both the Jewish and the Christian views of prophecy. For a different set of criteria by which to distinguish true from false prophecy see Jeremiah 23:19–22.

29. See Chapter 5, pp. 96–101.

30. On Amos 3:7, see n. 24 above.

31. Possibly the oracles against foreign nations are framed by this verse

that compares God's voice to the roaring of a lion (3:6) and God's roaring from Zion (1:2).

Chapter 2

1. Ivan Engnell, *Studies in Divine Kingship in the Ancient Near East*, p. 4.

2. George Steindorff and Keith C. Seele, *When Egypt Ruled the East*, pp. 82–86.

3. Henri Frankfort, *Kingship and the Gods: A Study of Ancient Near Eastern Religion and the Integration of Society and Nature*, p. 46.

4. The Amarna Letters are a body of diplomatic correspondence between officials in Palestine and Syria and the pharaohs Amenhotep III and Amenhotep IV (Akhenaton). They are dated the first quarter of the fourteenth century B.C.E.

5. David Lorton, "Toward a Constitutional Approach to Ancient Egyptian Kingship."

6. Engnell, *Studies in Divine Kingship*, pp. 5–6, 10.

7. James B. Pritchard, ed., *The Ancient Near East in Pictures Relating to the Old Testament*, figures 312–16.

8. Steindorff and Seele, *When Egypt Ruled the East*, pp. 85–86.

9. For a discussion of the relationship of the pharaoh to *ma'at*, see Keith W. Whitelam, *The Just King: Monarchical Judicial Authority in Ancient Israel*, pp. 26–28.

10. As quoted in Engnell, *Studies in Divine Kingship*, pp. 13–14.

11. Steindorff and Seele, *When Egypt Ruled the East*, pp. 85–86.

12. Frankfort, *Kingship and the Gods*, pp. 237–42. Thorkild Jacobsen, in *The Treasures of Darkness: A History of Mesopotamian Religion*, pp. 167–91, maintains the assembly of the gods in the *Enuma Elish* (the Babylonian Creation Epic), who elected Marduk as ruler over them, reveals a primitive form of democracy. The government was in the hands of an assembly of free men of the city who, in time of crisis, selected some individual to rule over them for a limited period. Jacobsen believes kingship gradually evolved from these ancient forms of democracy. See also Thorkild Jacobsen, "Primitive Democracy in Mesopotamia." On the relationship between the king and the gods in the Ancient Near East, see C. F. Gadd, *Idea of Divine Rule in the Ancient Near East*.

13. S. H. Hooke, *Babylonian and Assyrian Religion*, p. 27. Engnell, *Studies in Divine Kingship*, chapter II, believes such symbolism indicates the divine nature of Mesopotamian kings.

14. Possibly divine adoption is related to sacred marriage in Babylonian ritual: King Lipit–Ishtar was granted divine status by identifying himself with the god Urash in preparation for his marriage to the goddess Ishtar. See

Frankfort, *Kingship and the Gods*, pp. 297–98. The problem is that we don't know how "literally" to take the language. Undoubtedly there was a distinction between the metaphor of "divine sonship" and the empirical reality, but we cannot be sure what that distinction was.

15. Sigmund Mowinckel, *He That Cometh*, pp. 37–39.

16. For an English translation of *Enuma Elish* see "The Creation Epic," trans. E. A. Speiser, in James B. Pritchard, ed., *Ancient Near Eastern Texts Relating to the Old Testament*, 2nd ed., pp. 60–99, and "The Creation Epic: Additions to Tablets V–VII," trans. A. K. Grayson, in James B. Pritchard, ed., *Ancient Near Eastern Texts, Supplementary Texts and Pictures Relating to the Old Testament*, pp. 501–3.

17. In discussing the nation as the god's territory, G. W. Ahlström, *Royal Administration and National Religion in Ancient Palestine*, pp. 1–6, points out that in cities founded by Mesopotamian kings, two major houses were constructed: a house for the governor who represented the king, and a temple for the god, the ultimate ruler of the land. He writes: "The two buildings were the physical expressions of the national government representing king and god" (p. 2).

18. In like manner, Genesis 1:1–2:4a, the Priestly creation story, might better be called "the Exaltation of Yahweh," because its climax is not the creation of humankind but rather the establishment of the sabbath as the day when God is honored.

19. For a similar myth in which the young gods of order replace the older deities, see Hesiod's *Theogony*, in which Kronos defeats Ouranos and is in turn defeated by Zeus.

20. For concise descriptions of the major events in the *Akitu* Festival, see Baruch Halpern, *The Constitution of the Monarchy in Israel*, pp. 51–61; Mowinckel, *He That Cometh*, pp. 40–46; Frankfort, *Kingship and the Gods*, pp. 313–33.

21. On the relationship between the kings and temple building, see Arvid S. Kapelrud, "Temple Building, a Task for Gods and Kings."

22. Ahlström, *Royal Administration and National Religion in Ancient Palestine*, pp. 1–9, demonstrates that "the Mesopotamian king was, in principle, the organizer of the cult, the foundation the nation's life" (p. 9).

23. Some scholars interpret the humiliation of the king as a vestigial remains of a human sacrifice ritual in which a person is elected to serve as king for one day, at the close of which he is slain. No texts support such an interpretation. There is no emphasis on the death and resurrection of Marduk in the *Enuma Elish* or the *Akitu* Festival. The Mesopotamians did know of a substitute king who could replace the ruling monarch during a period of crisis, but it is pure conjecture that he was killed or sacrificed on behalf of the reigning monarch. Contrast with Engnell, *Studies in Divine Kingship*, chapter II.

24. A similar approach to myth and ritual is found in the Roman Catholic Mass. What takes place at the altar is regarded as a reflection of what is occurring in heaven (the cosmic realm). The earthly ritual does not make the sacrifice occur in heaven, but rather brings before the worshiper the true heavenly sacrifice.

25. *The Epic of Gilgamesh*, trans. N. K. Sandars, p. 68. It would seem that being two-thirds god and one-third man would qualify Gilgamesh for divinity. But the dramatic nature of the story rests on the mortal human nature of Gilgamesh, despite his divine longings.

26. The most famous of these law codes is the Code of Hammurabi. The stela that contains the law pictures Hammurabi receiving his laws from the sun god Shamash, the god of justice. For English text, see "The Code of Hammurabi," trans. Theophile J. Meek, in Pritchard, *Ancient Near Eastern Texts Relating to the Old Testament*, pp. 163–80. For a picture of the stela, see Pritchard, *The Ancient Near East in Pictures Relating to the Old Testament*, figure 246.

27. "Sumerian hymn to the king," trans. Hartmut Schmokel, in Walter Beyerlin, ed. *Near Eastern Religious Texts Relating to the Old Testament*, p. 107.

28. *The Epic of Gilgamesh*, trans. N. K. Sandars, p. 68.

29. "Lipit–Ishtar Lawcode," trans. S. N. Kramer, in Pritchard, *Ancient Near Eastern Texts Relating to the Old Testament*, p. 159.

30. Ahlström, *Royal Administration and National Religion in Ancient Palestine*, p. 7.

31. Mowinckel, *He That Cometh*, p. 48.

32. For English translations of these texts see Pritchard, *Ancient Near Eastern Texts Relating to the Old Testament*, Beyerlin, *Near Eastern Religious Texts*, and D. Winton Thomas, ed., *Documents from Old Testament Times*.

33. Baal is the most active god in Canaanite theology. He is also known as Hadad, the Amorite god of storm and rain clouds, "Him Who Mounts the Clouds," "Lord of the Plowed Furrows," represented in the form of a bull. Such imagery reveals his close relationship to the fertility of the land. His throne is located on Mount Saphon, thirty miles north of Ras Shamra.

34. As quoted in John Gray, *The Legacy of Canaan: The Ras Shamra Texts and Their Relevance to the Old Testament*, p. 56.

35. "Poems about Baal and Anath," trans. H. L. Ginsberg, in Pritchard, *Ancient Near Eastern Texts Relating to the Old Testament*, p. 140.

36. "The Legend of King Keret," trans. H. L. Ginsberg, in Pritchard, *Ancient Near Eastern Texts Relating to the Old Testament*, pp. 147–48.

37. "The Legend of King Keret," trans. John Gray, in Thomas, *Documents from Old Testament Times*, p. 121. The disease has made the king impotent.

38. Richard J. Clifford, "The Tent of El and the Israelite Tent of Meet-

ing"; see also his *The Cosmic Mountain in Canaan and the Old Testament*, pp. 123–31. Compare E. Theodore Mullen, Jr., *The Divine Council in Canaanite and Early Hebrew Literature*, pp. 168–75.

39. J. J. M. Roberts, "El," *IDB*, gen. ed. Keith Crim (Nashville: Abingdon Press, 1976), suppl. vol., 255–58. In support of Mount Amanus on the north Syrian coast for the location of El's tent, see Frank M. Cross, Jr., *Canaanite Myth and Hebrew Epic: Essays in the History of the Religion of Israel*, pp. 26–28, n. 50 on p. 161.

40. Tryggve N. D. Mettinger, "YHWH SABAOTH—The Heavenly King on the Cherubim Throne," pp. 109–38, figures 1, 2, 6.

Chapter 3

1. What is meant by state religion in the united kingdom of David and Solomon is the royal cultic theology and practice in Jerusalem. How widely this was known outside Jerusalem is difficult to determine. We do not know whether the common village peasant accepted this definition of Hebrew religion. But G. W. Ahlström, in *Royal Administration and National Religion in Ancient Palestine*, argues that through the establishment of cities and fortresses with cult centers state religion was disseminated throughout the nation.

2. On the relationship between dynastic kingship and a permanent sanctuary, see Moshe Weinfeld, "Zion and Jerusalem a Religious and Political Capital: Ideology and Utopia."

3. For a critique of classifying this material as history writing, see David M. Gunn, *The Story of King David: Genre and Interpretation*, pp. 20–21.

4. For a scholarly presentation of this interpretation of Israel's history, see John Bright, *A History of Israel*, pp. 107–43.

5. Laurence E. Stager, "The Archaeology of the Family in Ancient Israel."

6. This is the Albright theory of the Exodus. For evidence supporting this theory, see Bright, *A History of Israel*, pp. 120–33 and Bernhard W. Anderson, *Understanding the Old Testament*, pp. 46–52.

7. This theory was first proposed by George Mendenhall, "The Hebrew Conquest of Palestine," and developed in his *The Tenth Generation: The Origin of the Biblical Tradition*. See also his "Social Organization in Early Israel." Further development is found in C. H. J. de Geus, *The Tribes of Israel: An Investigation into some of the Presuppositions of Martin Noth's Amphictyony Hypothesis*, and Norman K. Gottwald's *The Tribes of Yahweh: A Sociology of the Religion of Liberated Israel, 1250–1050 B.C.E.* It forms the working hypothesis in two recently published survey books: Norman K. Gottwald's *The Hebrew Bible—A Socio-Literary Introduction* and J. Alberto

Soggins' *A History of Ancient Israel*. For a major criticism of this theory, see Jacob Milgrom, "Religious Conversion and the Revolt Model for the Formation of Israel." For a helpful clarification of the theory, see Gottwald's reply to his critics in "Two Models for the Origins of Ancient Israel: Social Revolution or Frontier Development."

8. Maxwell Miller and John H. Hayes, *A History of Ancient Israel and Judah*, pp. 54–121.

9. See Albrecht Alt, *Essays on Old Testament History and Religion*.

10. See J. Maxwell Miller, "The Israelite Occupation of Canaan," pp. 251–52.

11. For this analysis of the social structure of Israel during the settlement period, I am indebted to Miller and Hayes, *A History of Ancient Israel and Judah*, pp. 91–93. See also Robert R. Wilson, "Israel's Judicial System in the Preexilic Period," pp. 232–36, and de Geus, *Tribes of Israel*, pp. 133–56.

12. de Geus, *Tribes of Israel*, p. 118; Ahlström, *Royal Administration and National Religion in Ancient Palestine*, p. 33. Barnabas Lindars, in "The Israelite Tribes in Judah," maintains the twelve-tribe system originated during the reign of David and Solomon.

13. For a definition of these three tribal areas, see Miller and Hayes, *A History of Ancient Israel and Judah*, pp. 94–107.

14. See Miller and Hayes, *A History of Ancient Israel and Judah*, pp. 112–19, for a discussion of these sites and the various priestly lines who officiated at them. See also G. W. Ahlström, *Aspects of Syncretism in Israelite Religion*, pp. 14–24, 27–34.

15. The classical Canaanite El and Baal myths present Asherah as the consort of El and Anath as the consort of Baal. But the biblical material considers Asherah as Baal's consort. On the worship of Asherah and the meaning of the name, see Ahlström, *Aspects of Syncretism*, pp. 50–51.

16. On the identification of Yahweh with the various local manifestations of El, see Ahlström, *Aspects of Syncretism*, pp. 12–14.

17. Although proto-Israelite Yahwism, Yahwism of the settlement period, and that of the monarchy are probably different from one another, the nature of our sources makes it difficult to distinguish sharply among them.

18. Frank M. Cross, Jr., *Canaanite Myth and Hebrew Epic: Essays in the History of the Religion of Israel*, pp. 91–111.

19. Arvid S. Kapelrud, "Ugarit."

20. Patrick D. Miller, Jr., "El the Warrior," and his *The Divine Warrior in Early Israel*, pp. 48–58. Literature on the divine warrior motif is extensive. See Frank M. Cross, Jr. "The Divine Warrior in Israel's Early Cult"; Rudolf Smend, *Yahweh War and Tribal Confederation: Reflections upon Israel's Earliest History*; Peter C. Craigie, *The Problem of War in the Old Testament*;

Millar C. Lind, *Yahweh Is a Warrior: The Theology of Warfare in Ancient Israel*; John Day, *God's Conflict with the Dragon and the Sea: Echoes of a Canaanite Myth in the Old Testament*.

21. Harold W. Attridge and Robert A. Oden, Jr., eds., "Philo of Byblos: The Phoenician History," pp. 49, 51.

22. Trans. Cross, *Canaanite Myth and Hebrew Epic*, p. 101.

23. Ibid.

24. The Hebrew *yam sûp* should be translated Reed Sea. Red Sea comes from the Septuagint translation.

25. Trans. Cross, *Canaanite Myth and Hebrew Epic*, pp. 127–31.

26. These were the sacred lots borrowed from Canaanite culture. They may have been small pebbles or flat disks or dice used to divine the will of Yahweh.

27. The meaning of the ephod is even more obscure than the Urim and Thummim. Canaanite parallels indicate it may have been a robe worn by the priest that contained a pocket holding the Urim and Thummim. There is no evidence that either the ephod or the sacred lots were used after the reign of David.

28. For a good summary of the nature of Hebrew monarchy with emphasis on its place in the Hebrew culture, see Aubrey R. Johnson, "Hebrew Conceptions of Kingship."

29. Anointment was a ceremony in which oil was placed on the head, probably Egyptian in origin, but symbolic, in Hebrew faith, of the reception of God's spirit.

30. The Early Source refers to Saul as overseer, leader, prince (*nāgîd*), while the Late Source calls him king (*melek*).

31. See Georgio Bucellati, *Cities and Nations of Ancient Syria*.

32. Either Ish-baal, "man of Baal," or Esh-baal, "Baal exists" (I Chron. 8:33; 9:39). An editor altered the name to Ish-bosheth, "man of shame" (II Sam. 2:8) in criticism of Canaanite Baalism.

33. See R. N. Whybray, *The Succession Narrative: A Study of II Sam. 9–20; I Kings 1 and 2*.

34. Gunn, *Story of King David*.

35. See especially John Van Seters, *In Search of History: Historiography in the Ancient World and the Origins of Biblical History*, pp. 209–53, who argues that Israelite historiography developed under the influence of neo-Babylonian historiography in the late exilic period and reflects the faith and issues of that age. He believes the deuteronomistic history from Joshua through II Kings cannot be taken as a reliable source of information beyond, perhaps, the list of the kings of Israel and Judah. For a critique of this position, see Ziony Zevit, "Clio, I Presume," and "Deuteronomistic Historiography in 1 Kings 12–2 Kings 17 and the Reinvestiture of the Israelian Cult."

36. Soggins, in his recent book *A History of Ancient Israel*, p. 26, argues that the history of Israel begins with the united kingdom under David and Solomon.

37. Miller and Hayes, *A History of Ancient Israel and Judah*, pp. 149–88, limits considerably the size of David's empire.

38. Kingship in the Ancient Near East at this time was hereditary. It has been argued that both Saul and David ruled by popular recognition of their charismatic powers, and that this tradition prevailed in the northern kingdom of Israel after the division. However, it appears that all kings of both Judah and Israel, including Saul and David, desired to pass on their thrones to their heirs. Only Judah succeeded in establishing a permanent dynasty.

39. Concerning the rivalry between the Saulides and David, see James W. Flanagan, "Chiefs in Israel."

40. Concerning David's legal claim on Michal, see Zafrira Ben-Barak, "The Legal Background to the Restoration of Michal to David."

41. J. J. M. Roberts, "The Davidic Origin of the Zion Tradition."

42. Gottwald, *Tribes of Yahweh*, p. 204, suggests that the conflict between Canaanite and Israelite cultures may have begun when David absorbed the Canaanite population into monarchical Israel. Before this there may have been a period of peaceful coexistence. With David's reign, Yahwism defined itself over against Canaanite culture.

43. II Samuel 24:6–7 lists Sidon and Tyre as part of David's empire included in the census. According to I Kings 11:1, one of Solomon's wives was from Sidon.

44. Gottwald, *Tribes of Yahweh*, pp. 368–69.

45. For a list of the high officials of David and Solomon and a detailed discussion of their functions, see Tryggve N. T. Mettinger, *Solomonic State Officials: A Study of the Civil Government Officials of the Israelite Monarchy*. See also the chapter on "The New Bureaucracy" in E. W. Heaton, *Solomon's New Men: The Emergence of Ancient Israel as a National State*, pp. 47–60.

46. Miller and Hayes, *A History of Ancient Israel and Judah*, pp. 113, 172, 186. See also Ahlström, *Royal Administration and National Religion in Ancient Palestine*, pp. 47–51.

47. On the land grant policy of the monarchy, see Mettinger, *Solomonic State Officials*, pp. 80–85.

48. Miller and Hayes, *A History of Ancient Israel and Judah*, map 15, p. 181. Ahlström, *Royal Administration and National Religion in Ancient Palestine*, pp. 51–65, believes the list of Levitical cities derives from the postexilic period, although he does argue that cities were used as administrative and religious centers during the monarchy.

49. Ahlström, *Royal Administration and National Religion in Ancient Palestine*, p. 8.

50. If one accepts the position that Israel emerged primarily from *within* the land, then possibly David was attempting to distinguish between Judean and Canaanite practices.

51. For a discussion of the ark, Zion, and the temple as part of a national religion, see Tomoo Ishida, *The Royal Dynasties in Ancient Israel: A Study on the Formation and Development of Royal-Dynastic Ideology*, pp. 136–50.

52. J (Yahwistic), E (Elohistic), D (Deuteronomic), and P (Priestly) refer to four literary or oral sources of the pentateuch. The most widely accepted dates for these traditions are 950 B.C.E. (J), 850 (E), 550 (D), and 450 (P).

53. See G. Henton Davies, "Ark of the Covenant."

54. See p. 27.

55. For an excellent discussion of ark and cherubim, see R. E. Clements, *God and Temple*, pp. 28–35, who argues that the Shiloh sanctuary was dedicated to a Canaanite god, El *Ṣĕbāʾôt* (God of hosts), who was regarded as a divine king, enthroned on the cherubim. When the ark was placed there, it was the first time Yahweh was understood as a divine king over Israel. See also Cross, *Canaanite Myth and Hebrew Epic*, p. 69.

56. J. R. Porter, "The Interpretation of 2 Samuel VI and Psalm CXXXII."

57. Francis Brown, S. R. Driver, and Charles A. Briggs, *Hebrew and English Lexicon of the Old Testament*, S. V. "*ṣĕbāʾôt*," p. 839; J. Obermann, "The Divine Name YHWH in the Light of Recent Discoveries"; Bernhard W. Anderson, "Host, Host of Heaven"; Ludwig Köhler, *Old Testament Theology*, pp. 49–51.

58. S. R. Driver, S. V. "Lord of Hosts."

59. J. P. Ross, "Yahweh *Ṣĕbāʾôt* in Samuel and Psalms." See p. 90, note 1, for the reasons Ross favors Baal *Ṣĕbāʾôt* to El *Ṣĕbāʾôt* for the origin of the "Lord of hosts."

60. Nehushtan, the bronze serpent, was also a temple cult object. It was probably a Baal fertility symbol, part of the Jebusite worship center in pre-Israelite Jerusalem. It may have been the name of a deity, for the mother of King Jehoiakim was called Nehushta (II Kings 24:8). Although the cherubim, also Canaanite in origin, were compatible with Yahweh worship, the bronze serpent, similar to the bull images in the north, had fertility cult qualities that conflicted with orthodox Yahwism. See H. H. Rowley, "Zadok and Nehushtan." It was destroyed in the religious reforms of Hezekiah (II Kings 18:4). Numbers 21:8–9 is undoubtedly an etiological story tracing its origin to the Mosaic period.

61. In the Priestly account (Exod. 25–30, 35–40) the ark, overlaid with gold, is placed in an elaborate tabernacle. There is little doubt that this represents a late tradition.

62. In private correspondence, E. Theodore Mullen, Jr. points out that in both II Samuel 6:17 and 7:12, passages relating David's placing of the ark in Jerusalem, only "the tent," *not* "the tent of meeting," is mentioned. It is not until I Kings 8:14, at Solomon's dedication of the temple, that the ark and the tent of meeting are joined.

63. For the identification of Yahweh with El at Shiloh, see Ishida, *Royal Dynasties in Ancient Israel*, p. 39.

64. For a summary of Zion theology, see Aubrey R. Johnson, *Sacral Kingship in Ancient Israel*, pp. 31–53.

65. For example, see John H. Hayes, "The Tradition of Zion's Inviolability."

66. Roberts, "Davidic Origin of the Zion Tradition"; see also his articles "Zion Tradition" and "Zion in the Theology of the Davidic–Solomonic Empire."

67. Trans. Roberts, "Davidic Origin of the Zion Tradition," p. 334.

68. See p. 26.

69. Contrast with J. W. McKay, *Religion in Judah under the Assyrians, 732–609 B.C.*

70. Hebrew poetry unites mythology and history. In Psalm 74, the plea for deliverance from the enemy is related to God's triumph over the cosmic forces at creation. Though postexilic in its present form, this psalm preserves preexilic imagery.

71. While no siege of Jerusalem is mentioned in the text, the valley of Rephaim is probably located southwest of Jerusalem. However, to interpret II Samuel 5:24a ("for then the LORD has gone out before you to smite the army of the Philistines") within the Zion tradition seems rather forced.

72. The issues regarding the dating of and the relationship among various passages referring to the Davidic covenant are complex. I have followed the moderate position of Moshe Weinfeld. For evidence supporting a later preexilic and exilic date for the Davidic covenant, see Tryggve N. D. Mettinger, *King and Messiah: The Civil and Sacral Legitimation of the Israelite Kings*, pp. 254–93. Mettinger argues that Israelite royal ideology moved from an early interest in the divine sonship of the kings (Pss. 2 and 110) to a later development of the Davidic dynastic promise. Furthermore, in the early preexilic period this Davidic dynastic promise took the form of an unconditional covenant. See Mettinger's analysis of the various layers found in II Samuel 7 and Psalm 89.

73. Moshe Weinfeld, "The Covenant of Grant in the OT and in the Ancient Near East," *Deuteronomy and the Deuteronomic School*, pp. 59–116; "Covenant, Davidic."

74. While the wording of the prophecy betrays the hand of the deuteronomistic historians, the essential content comes from the period of the early

monarchy. The final version of the deuteronomistic history was written after the fall of the Davidic dynasty in 587; the writers probably expected a restoration of the Davidic line in the near future.

75. As quoted in Weinfeld, "Covenant, Davidic," p. 190.

76. As quoted in ibid.

77. The covenant concept is conveyed through a variety of terms: *bĕrît*, "covenant" (Ps. 89:3 [H 4], 28 [H 29], 34 [H 35], 39 [H 40]); *ḥesed*, "steadfast loyalty" (II Sam. 7:15; Ps. 89:1 [H 2], 2 [H 3], 14 [H 15], 19 [H 20], 24 [H 25], 28 [H 29], 33 [H 34], 49 [H 50); *ṭôbâ*, "favor" (I Sam. 25:30; II Sam. 7:28); *dābār*, "word" (II Sam. 7:25–28); *'ĕmûnâ*, "faithfulness" (Ps. 89:1 [H 2], 2 [H 3], 5 [H 6], 8 [H 9], 24 [H 25], 33 [H 34], 49 [H 50]. Similar terminology is found in Assyrian and Hittite royal grants.

78. When applied to a human being, the term "son of God" means a person is godly. Similarly, "sons of prophets" designates persons who are propheticlike in their behavior, "son of man" is a synonym for a human being, and "son of righteousness" means one who is righteous.

79. Cross, *Canaanite Myth and Hebrew Epic*, p. 233, argues that Psalm 132 comes from the early Jerusalem cultus at the beginning of David's reign when kingship was considered conditional. He believes the unconditional nature of the Davidic covenant in II Samuel 7:12–16 and Psalm 89 reflects the post-Davidic Judean ideology of kingship. The divine-adoption motif he attributes to Canaanite ideology.

80. For example, see R. E. Clements, *Abraham and David: Genesis XV and Its Meaning in Israelite Tradition*.

81. Roland de Vaux, *Ancient Israel: Its Life and Institutions*, pp. 113–14.

82. Chapter 6 is devoted to a fuller discussion of the judicial system in Ancient Israel.

83. We know little about early Israelite legal procedures. Certainly the legalism of the elders at the gates would not be sufficient for the administration of an empire. See Robert R. Wilson, "Israel's Judicial System in the Preexilic Period," for a helpful survey of what can be learned from the biblical texts and comparative sociological studies.

84. On the possible relationship of David's ecstatic dance and the *hieros gamos*, see Ahlström, *Aspects of Syncretism*, pp. 35–37.

85. Miller and Hayes, *A History of Ancient Israel and Judah*, p. 173.

86. See Rowley, "Zadok and Nehushtan." Contrast with de Vaux, *Ancient Israel*, pp. 114, 374. See also Ahlström, *Royal Administration and National Religion in Ancient Palestine*, pp. 29–30, esp. n. 15. G. Widengren, *Sacrales Königtum im Alten Testament und im Judentum*, p. 47, believes El Elyon was the chief god in pre-Davidic Jerusalem.

87. See Rowley, "Zadok and Nehushtan," and J. R. Bartlett, "Zadok and His Successors at Jerusalem."

88. Cross, *Canaanite Myth and Hebrew Epic*, pp. 207–15.

89. There is no evidence to support Johnson's claim in "The Role of the King in the Jerusalem Cultus," p. 83, that Solomon's temple was built on the site of an earlier sun worship of Elyon. On the solar cult, see F. J. Hollis, "The Sun Cult and the Temple at Jerusalem." For text and interpretation of Shalem, see J. C. L. Gibson, *Canaanite Myths and Legends*, pp. 28–30, 123–27. See also John Gray, *The Legacy of Canaan*: pp. 170–174. and R. A. Rosenberg, "Shalem (God)."

90. For further details on Canaanite influence on Solomon's temple, see Ahlström, *Aspects of Syncretism*, pp. 43–46. In a later publication, Ahlström, *Royal Administration and National Religion in Ancient Palestine*, pp. 35–36, suggests that Solomon's temple was either patterned after Egyptian models, or it was possibly a unique structure, making it an Israelite contribution to the architecture of the Ancient Near East. On the similarity between Solomon's temple construction and temple building by kings throughout the Ancient Near East, see Arvid S. Kapelrud, "Temple Building, a Task for Gods and Kings." For Israelite opposition to the temple and support for the tent tradition, see Virgil W. Rabe, "Israelite Opposition to the Temple."

91. Following Eissfeldt, Artur Weiser, in *The Psalms, A Commentary*, pp. 539–40, believes the psalm has its origin in a cultic tradition prior to the division of the kingdom. It reflects the incorporation of the Davidic tradition into the Jerusalem temple cult. The wording of v. 67 does not refer to the destruction of the northern kingdom of Israel, but to the choice of David from Judah as ruler over the united kingdom. Because it contains no reference to the destruction of the Jerusalem temple, Weiser takes this as proof of its preexilic date.

92. As Zevit points out in "Deuteronomistic Historiography in 1 Kings 12–2 Kings 17 and the Reinvestiture of the Israelian Cult," that the northern kings were simply doing what was within the perogative of all kings. He writes on p. 61:

> Both biblical and extra-biblical evidence indicates that the types of activities condemned in Jeroboam were within the recognized traditional authority of kings. From sources presented by Dtr himself, it may be learned that kings did appoint non-Levitical priests (2 Sam. 8:18), did dismiss Levitical priests (1 Kgs. 2:26), did build temples and shrines (1 Kgs. 6; 16. 32–33; 2 Kgs. 21.2–7), and in general were involved in cultic policies and politics (2 Kgs. 16, 2–4, 8, 10–18; 18:2–5; 10:11; 23:4–20).

93. J. P. J. Oliver, "In Search of a Capital for the Northern Kingdom," argues that Israel's first capital was Samaria. He maintains that before Omri's reign the nation was too disorganized to have a capital. Tirzah, Shechem, and Penuel were merely headquarters for the royalty.

94. On the place of Bethel in Jeroboam I's administration of Israel, see

Ahlström, *Royal Administration and National Religion in Ancient Palestine*, pp. 58–61.

95. On the other hand, possibly Jeroboam I was simply reviving worship at the Canaanite shrines of Bethel and Dan. In this case, he would be abolishing the distinction made by David between Judean and Canaanite practices. This is, at least, the basis for the deuteronomistic historians' attack on Jeroboam I.

96. Contrast the Book of Judges that reflects the tradition that the priests of Dan and Bethel were of the Levitical line, Bethel tracing its priesthood back to Aaron (Judg. 20:26–28) and Dan tracing its priesthood back to Moses (Judg. 18:30). The situation is further complicated by the obvious parallels between Jeroboam's erection of the golden calves and the apostasy of Aaron in the wilderness (see Exod. 32). Possibly, Jeroboam was deliberately imitating Aaron in order to help legitimate Yahweh worship in the north. More likely, however, Exodus 32 reflects Zadokite opposition to the bull cult in the north. See Moses Aberbach and Leivy Smolar, "Aaron, Jeroboam, and the Golden Calves."

97. On the cult and calendar established by Jeroboam I, see Julian Morgenstern, "Three Calendars of Ancient Israel"; William F. Albright, *From the Stone Age to Christianity*, pp. 298–301; Aberbach and Smolar, "Aaron, Jeroboam, and the Golden Calves"; John Gray, *I and II Kings, A Commentary*, pp. 311–39; de Vaux, *Ancient Israel*, pp. 332–36; S. Talmon, "Divergences in Calendar-Reckoning in Ephraim and Judah."

Chapter 4

1. Oracles against Damascus (1:3–5), Philistia (1:6–8), Tyre (1:9–10), Edom (1:11–12), Ammon (1:13–15), Moab (2:1–3). Other literary units containing a series of oracles against foreign nations are Isaiah 13–23, Jeremiah 46–51, and Ezekiel 25–32. For a study of these and related texts, see Duane L. Christensen, *Transformations of the War Oracles in Old Testament Prophecy: Studies in the Oracles against the Nations*.

2. James L. Mays, *Amos, A Commentary*, pp. 41–42; James M. Ward, *Amos & Isaiah: Prophets of the Word of God*, pp. 69 and 99, n. 4; Hans Walter Wolff, *Joel and Amos*, pp. 163–64. For arguments defending its authenticity, consult Erling Hammershaimb, *The Book of Amos, A Commentary*, pp. 43–46. William Rainey Harper, *A Critical and Exegetical Commentary on Amos and Hosea*, pp. 44–47, presents evidence for and against its genuineness, drawing no conclusion. Richard S. Cripps, *A Critical & Exegetical Commentary on the Book of Amos*, pp. 137–39, does not raise the issue.

3. For arguments equating the "lies" (*kāzāb*) of this passage with "idols" (*hebel*) in the deuteronomistic style (Deut. 32:21; I Kings 6:13, 26; II Kings 17:15), see Wolff, *Joel and Amos*, p. 164, and Mays, *Amos*, pp. 41–42.

4. Compare "to keep his statutes" (Amos 2:4) with Deuteronomy 4:5–6, 40; 5:1; 6:17; 7:11; 11:32; 17:19; 26:16–17. See also the same style in I Kings 3:14; 8:58; 9:4; II Kings 17:37; 23:3.

5. See Paul J. Achtemeier, *The Inspiration of Scripture: Problems and Proposals*, for an excellent argument employing the results of redaction criticism in defense of the authority of Scripture.

6. Harper, *Amos and Hosea*, pp. 27–34; Mays, *Amos*, pp. 33–36; Wolff, *Joel and Amos*, pp. 140, 158–60. Those who accept these oracles as authentic Amos are Cripps, *Book of Amos*, pp. 126–32; Ward, *Amos & Isaiah*, p. 99; and Hammershaimb, *Book of Amos*, pp. 32–38.

7. It should be noted, however, that a characteristic of oral poetry is a highly structured system of repetitive parallelism.

8. See Obadiah; Lamentations 4:21–22; Isaiah 34:5; Jeremiah 49:7–22; Ezekiel 25:12–14; Joel 3:19; Psalm 137:7.

9. See Joel 3:6 and Ezekiel 27:13.

10. Christensen, *Transformations of the War Oracles in Old Testament Prophecy*, pp. 17–72.

11. As reconstructed and translated by Christensen, *Transformations of the War Oracles in Old Testament Prophecy*, pp. 24–25.

12. Wolff, *Joel and Amos*, pp. 137–38, claims the graduated numbers reflect a wisdom tradition. However, the use of numerical parallelism is characteristic of northwest Semitic poetry. See S. Gevirtz, *Patterns in the Early Forms of Prophetic Speech*, pp. 15–24.

13. See also the Septuagint reading of Deuteronomy 32:8–9, in which Elyon assigns the nations to various deities; Yahweh receives Israel.

> When the Most High (Elyon) gave to the
> nations their inheritance,
> when he separated the sons of men,
> he fixed the bounds of the peoples according
> to the number of the sons of God.
> For the LORD's portion is his people,
> Jacob his allotted heritage.

14. Naaman's actions are similar to Jonah's, the narrow-minded prophet who was the central character of the postexilic didactic story that bears his name. Jonah confesses that he fears the Lord who is the creator of the sea and dry land (Jon. 1:9), but attempts to escape God's presence by sailing toward Tarshish (1:3).

15. Norman K. Gottwald, *All the Kingdoms of the Earth: Israelite Prophecy and International Relations in the Ancient Near East*, pp. 94–119; Arvid S. Kapelrud, *Central Ideas in Amos*, pp. 33–59.

16. Gottwald, *All the Kingdoms of the Earth*, pp. 103–12.

17. Ibid., pp. 110–12.

18. Gottwald adopts the Septuagint reading. Wolff, *Joel and Amos*, pp. 189 and 192–93, supports reading "Ashdod," following the Masoretic Hebrew text.

19. See Wolff, *Joel and Amos*, pp. 274–75.

20. Gottwald, *All the Kingdoms of the Earth*, p. 112.

21. See Cripps, *Book of Amos*, pp. 262–64; Mays, *Amos*, pp. 156–60; Ward, *Amos & Isaiah*, pp. 72–73.

22. Kapelrud, *Central Ideas in Amos*, p. 29.

23. See Frank M. Cross, Jr. *Canaanite Myth and Hebrew Epic: Essays in the History of the Religion of Israel*, part I.

24. Yet it can be argued that attributing El imagery to Yahweh probably influenced the form Hebrew monotheism took.

25. John Barton, *Amos's Oracles against the Nations*, p. 2. See also G. W. Ahlström, *Royal Administration and National Religion in Ancient Palestine*, p. 18, n. 50.

26. Barton, *Amos's Oracles against the Nations*, pp. 51–61.

27. Aage Bentzen, "The Ritual Background of Amos i 2–ii 6."

28. Kapelrud, *Central Ideas in Amos*, pp. 17–33; Gottwald, *All the Kingdoms of the Earth*, pp. 103–12.

29. M. Weiss, "The Pattern of the 'Execration Texts' in The Prophetic Literature." Wolff, *Joel and Amos*, pp. 145–46, lists five objections to Bentzen's theory.

30. Aside from geographic patterns, the Egyptian Execration Texts also indicate that the names of the nations and the curses directed against them are pronounced. This may be related to the prophetic oracle of judgment.

31. See Chapter 3, pp. 46–48.

32. John Mauchline, "Implicit Signs of a Persistent Belief in the Davidic Empire."

33. Ibid., p. 289.

34. While these two passages (9:7 and 8b–12) are in the vision section of the Book of Amos, they are thematically related to the oracles against foreign nations. They probably owe their present position in the text to the final redactor. Because the first is a doom oracle and the second a hope oracle, they formed an appropriate summary of Amos' message.

35. Amos 9:5–6 is one of the three doxologies in the book that most scholars believe betray the hand of a redactor.

36. On the origin of the Philistines, see Trude Dothan, *The Philistines and Their Material Culture*, pp. 21–23.

37. Amos apparently depended on Judean traditions concerning the giant stature of the Amorites who occupied the land before it was taken by Israel. See Martin Noth, *Numbers, A Commentary*, on Numbers 13:28.

38. An even more negative approach to the Ethiopians is offered by Harper, *Amos and Hosea*, p. 192.

39. Francis Brown, S. R. Driver, and Charles A. Briggs, *A Hebrew and English Lexicon of the Old Testament*, S. V. *"peša'*," p. 833a. See also Klaus Koch, *The Prophets, Volume One, The Assyrian Period*, pp. 61–62, and Eryl W. Davies, *Prophecy and Ethics: Isaiah and the Ethical Traditions of Israel*, pp. 45–48. Ahlström, *Royal Administration and National Religion in Ancient Palestine*, p. 3, n. 12, points out that in the Ancient Near East a rebellion against the king, who is the divine representative, is also an act of insubordination against the god.

40. It is true that these atrocities are comparable to the deuteronomic ideal of the *ḥerem*. This reveals once again the nationalistic nature of the biblical material. There is, alas, a tendency in most peoples to condemn as barbaric in others certain practices in which they themselves engage.

41. Amos 9:13–15 is generally regarded as a highly symbolic postexilic hope addition almost apocalyptic in tone. See Wolff, *Joel and Amos*, pp. 350–55; Mays, *Amos*, pp. 165–68.

42. Wolff, *Joel and Amos*, pp. 344–49; Mays, *Amos,* p. 160–65; Ward, *Amos & Isaiah*, pp. 87–90.

43. Wolff, *Joel and Amos*, p. 348.

44. Ibid., p. 353.

45. Mays, *Amos*, p. 164.

46. Hammershaimb, *Book of Amos*, p. 135.

47. Wolff, *Joel and Amos*, p. 353.

48. Harper, *Amos and Hosea*, p. 198.

49. Cripps, *Book of Amos*, p. 320.

50. Mauchline, "Implicit Signs," p. 291.

51. H. Neil Richardson, "SKT (Amos 9:11): 'Booth' or 'Succoth'?" Concerning variations on the Hebrew spelling of Succoth, see p. 377.

52. Yigael Yadin, *The Art of Warfare in Biblical Lands*, pp. 270–75, 305–10. On the place of Succoth in the Davidic kingdom, see Norman K. Gottwald, *The Tribes of Yahweh: A Sociology of the Religion of Liberated Israel, 1250–1050 B.C.E.*, pp. 573–77.

53. Against Yadin, it could be argued that if the ark was understood to have been housed in a tent, it is conceivable to place it in a booth. Furthermore, it was quite common for the auxiliary troops to have better accommodations than those engaged in battle.

54. Yadin, *Art of Warfare*, pp. 305–10.

55. Richardson, "SKT (Amos 9:11): 'Booth' or 'Succoth'?", p. 301.

56. In support of this translation of the Masoretic text and its relationship to royal theology, see John Hayes, "The Usage of Oracles against Foreign Nations in Ancient Israel, " p. 91.

57. John Bright, *A History of Israel*, pp. 273–75.

58. See Chapter 1, p. 8.

59. The word fire appears nine times in the Book of Amos, always as a

feminine noun. However, possibly these oracles were uttered in a fixed ritual ceremony in which the antecedent of the pronoun was known.

60. On reading "Assyria" in Amos 3:9, see n. 18 of this chapter.

61. For this approach to the visions, see Mays, *Amos*, pp. 123–30. Chapter 7, pp. 156–60 presents a fuller discussion of these visions.

62. On divine fire in the Ancient Near East, see Patrick D. Miller, Jr., "Fire in the Mythology of Canaan and Israel," and Delbert R. Hillers, "Amos 7,4 and Ancient Parallels."

63. J. Maxwell Miller, "The Elisha Cycle and the Accounts of the Omride Wars."

64. The Assyrians called him Adad-idri. The biblical material gives him the same name as the son of Hazael, Ben-hadad.

65. As in Bright, *A History of Israel*, pp. 234–56.

66. Barton, *Amos's Oracles against the Nations*, pp. 27–31.

67. Menahem Haran, "The Rise and Decline of the Empire of Jeroboam ben Joash."

68. Wolff, *Joel and Amos*, pp. 288–89.

69. Ibid., p. 158.

70. Menahem Haran, "Observations on the Historical Background of Amos 1:2–2:6," presents historical arguments for emending 1:6, 9 to read "Aram" instead of "Edom."

71. Although the punishments are the same, variation is seen in the arrangement of vv. 5 and 8:

A	I will break the bar of Damascus,
B	and cut off the inhabitants from the Valley of Aven,
C	and him that holds the scepter from Beth-eden;
D	and the people of Syria shall go into exile to Kir.

(v.5)

B	I will cut off the inhabitants from Ashdod,
C	and him that holds the scepter from Ashkelon;
A	I will turn my hand against Ekron;
D	and the remnant of the Philistines shall perish.

(v.8)

72. Wolff, *Joel and Amos*, pp. 158–60; Mays, *Amos*, pp. 33–34.

73. Wolff, *Joel and Amos*, p. 159.

74. Among those scholars who regard the oracle against Tyre as genuine Amos are Harper, *Amos and Hosea*, pp. 28–31; Cripps, *Book of Amos*, pp. 126–29; Hammershaimb, *Book of Amos*, pp. 34–35; Bentzen, "Ritual Background of Amos i 2–ii 6."

75. It is possible that Tyre and Israel formed a joint maritime venture. See S. Yevin, "Did the Kingdom of Israel have a Maritime Policy?"

76. F. C. Fensham, "The treaty between the Israelites and Tyrians," believes both David and Solomon formed parity treaties with the Tyrians. He regards the biblical references in II Samuel and I Kings as historically reliable.

77. See evidence for the covenant basis of this language under discussion of the oracle against Edom that follows.

78. Some scholars regard the abbreviated judgment passage as evidence that the oracle is not genuine Amos.

79. Wolff, *Joel and Amos*, p. 160; Mays, *Amos*, pp. 35–36; Harper, *Amos and Hosea*, pp. 31–34.

80. See also Isaiah 34:5–14; Jeremiah 49:7–22; Joel 3:19; Psalm 137:7.

81. See Michael Fishbane, "The Treaty Background of Amos 1:11 and Related Matters"; John Priest, "The Covenant of Brothers"; Haran, "Observations on the Historical Background of Amos 1:2–2:6."

82. For a critique of Fishbane's argument, see Robert B. Coote, "Amos 1:11: RHMYW." See Fishbane's reply, "Additional Remarks on RHMYW (Amos 1:11)."

83. Keith N. Schoville, "A note on the oracles of Amos against Gaza, Tyre, and Edom."

84. "The Moabite Stone," trans. E. Ullendorff, in D. Winton Thomas, ed., *Documents From Old Testament Times*, pp. 195–98; "The Mesha Inscription," trans. E. Lipinski, in Walter Beyerlin, ed., *Near Eastern Religious Texts Relating to the Old Testament*, pp. 237–40; "The Moabite Stone," trans. W. F. Albright, in James B. Pritchard, ed., *Ancient Near Eastern Texts Relating to the Old Testament*, pp. 320–21.

Chapter 5

1. The exceptions are Elah (I Kings 16:8–14) and Shallum (II Kings 15:13–15).

2. Frank E. Eakin, Jr., "Yahwism and Baalism before the Exile."

3. Morton Cogan, *Imperialism and Religion: Assyria, Judah, and Israel in the Eighth and Seventh Centuries B.C.E.*

4. See Gunnar Östborn, *Yahweh and Baal: Studies in the Book of Hosea and Related Documents.*

5. For a critique of an overreliance on redaction criticism, see Hans M. Barstad, *The Religious Polemics of Amos.*

6. C. R. North, "Pentateuchal Criticism," p. 56.

7. For a balanced discussion of this issue, see C. F. Whitley, *The Prophetic Achievement*, pp. 63–92.

8. See E. W. Heaton, *The Old Testament Prophets*, pp. 61–65; John Skinner, *Prophecy and Religion*, pp. 181–82. By cultic religion I mean the ritual practices occurring at sanctuaries where officially approved priesthoods function.

9. H. H. Rowley, "The Unity of the Old Testament," pp. 347–48; H. Wheeler Robinson, *Inspiration and Revelation in the Old Testament*, p. 226; A. C. Welch, *Prophet and Priest in Old Israel*, p. 21.

10. If the internal revolt is accepted, the Israelites would simply have rededicated to the new god, Yahweh, the shrines at which they had previously been worshiping.

11. Hans Walter Wolff, *Joel and Amos*, pp. 264–65; Hans-Joachim Kraus, *Worship in Israel: A Cultic History of the Old Testament*, pp. 112–13.

12. The meaning of Hosea 2:16 is often lost in translation. The term "*baʿal*" means master or lord. It was used by a devotee to refer to the Canaanite deity (and by the Hebrew worshiper to refer to Yahweh), by a slave to refer to a master, and by a wife to refer to her husband. Hosea condemned its usage and anticipated a time when God would be called by the affectionate *ʾîšî*, "my husband."

13. For example, Wolff, *Joel and Amos*, pp. 166–68, 276–78; James L. Mays, *Amos, a Commentary*, pp. 46–50, 116–17.

14. See Marvin H. Pope, "A Divine Banquet at Ugarit"; Barstad, *Religious Polemics of Amos*, pp. 135–38; Robert B. Coote, *Amos among the Prophets: Composition and Theology*, pp. 37–39.

15. Pope, "Divine Banquet at Ugarit," p. 193.

16. Barstad, *Religious Polemics of Amos*, p. 127, n. 5, holds that the bowl was used in cultic connections and does not refer to drinking wine in abundance.

17. For a description of ivory beds used in these rites, see Hershel Shanks, "Ancient Ivory: The Story of Wealth, Decadence, and Beauty," p. 45.

18. On the relationship between love and death in funeral feasts celebrated with wine, women, and song in the Ancient Near East, see Marvin H. Pope, *Song of Songs*, pp. 210–29.

19. Barstad, *Religious Polemics of Amos*, pp. 11–36.

20. Ibid., p. 15.

21. See Wolff, *Joel and Amos*, pp. 166–68; Mays, *Amos*, pp. 46–50. Some older commentaries relate the text to cultic prostitution: see Richard S. Cripps, *A Critical & Exegetical Commentary on the Book of Amos*, pp. 141–43; William Rainey Harper, *A Critical and Exegetical Commentary On Amos and Hosea*, pp. 51–52.

22. Cripps, *Book of Amos*, p. 165; Harper, *Amos and Hosea*, p. 86; Wolff, *Joel and Amos*, pp. 205–6; Mays, *Amos*, p. 71–73.

23. Mays, *Amos*, p. 72.

24. For the placing of representations of a deity upon standards, see James B. Pritchard, ed., *The Ancient Near East in Pictures Relating to the Old Testament*, figures. 305, 625, 684.

25. See p. 87, n. 11.

26. Wolff, *Joel and Amos*, pp. 288–89.

27. Menahem Haran, "The Rise and Decline of the Empire of Jeroboam ben Joash," p. 272.

28. H. H. Rowley, "The Samaritan Schism in Legend and History."

29. Shalom M. Paul, "Sargon's Administrative Diction in II Kings 17, 27."

30. Although the deuteronomists mention only the golden calves at Bethel and Dan, these images may well have existed at other cultic centers.

31. This is similar to the deuteronomists' substitution of *bō'-šet* ("shame") for baal in Hebrew names. Ish-baal ("man of the Lord") or Esh-baal ("the Lord exists") in I Chronicles 8:33; 9:39 is Ish-bosheth ("man of shame") in II Samuel 2:8; Meribaal ("he who strives for the Lord") in I Chronicles 8:34; 9:40 is Mephibosheth ("he who spreads shame") in II Samuel 4:4; 9:6–13.

32. Barstad, *Religious Polemics of Amos*, pp. 159–85.

33. Ibid., p. 187.

34. Erling Hammershaimb, *The Book of Amos, A Commentary*, pp. 129–30.

35. John D. W. Watts, *Vision and Prophecy in Amos*, p. 44, n. 12.

36. Arvid S. Kapelrud, *Central Ideas in Amos*, p. 50.

37. Frank J. Neuberg, "An Unrecognized Meaning of Hebrew DOR." See also Peter R. Ackroyd, "The Meaning of Hebrew *dwr* Reconsidered."

38. Barstad, *Religious Polemics of Amos*, pp. 193–98. See also Mays, *Amos*, p. 150.

39. Some of Amos' criticism of the cult may have been delivered at Bethel without mentioning the center by name. For example, 2:7b–8; 5:21–24; 9:1–4.

40. See later for discussion of Beer-sheba as a northern pilgrim site located within Judah.

41. Wolff, *Joel and Amos*, pp. 163–64.

42. Artur Weiser, *Die Profetie des Amos*, pp. 229–31.

43. Wolff, *Joel and Amos*, pp. 270, 274–75.

44. Ibid., p. 175.

45. Although Hosea 8:4; 9:9; 10:9; and 13:11 may indicate an antimonarchical position.

46. There were probably a number of additional worship centers in Israel beyond those mentioned in the Book of Amos. A Hebrew seal, which is said to have been found near ancient Samaria, is inscribed: ["Belonging to Ze]kharyau, priest of Dor." It is dated ca. eight century B.C.E., indicating that a temple existed at this administrative site (I Kings 4:11) in the central region near the coast. See N. Avigad, "The Priest of Dor." The Book of Amos does not mention Dor.

47. Coote, *Amos among the Prophets*, develops Wolff's Bethel editor hypothesis in the extreme, leaving no references to sanctuaries as recognizable at Coote's A-stage level of Amos.

48. Wolff, *Joel and Amos*, p. 200.

49. Ibid., p. 239, argues that the Beer-sheba reference is a later addition to be ascribed to the disciples of Amos in Judah. Contrast A. Vanlier Hunter, *Seek the Lord! A Study of the Meaning and Function of the Exhortation in Amos, Hosea, Isaiah, Micah, and Zephaniah*, p. 69, who argues that the sequence "Bethel–Gilgal–Beersheba–Gilgal–Bethel" forms a perfect chiasma with Beer-sheba the middle component.

50. The Bethel editor probably thought the north worshiped Baal at Bethel; certainly the deuteronomists did.

51. Wolff, *Joel and Amos*, pp. 220–25, maintains 4:6–12 is from the hand of a redactor.

52. Ibid., pp. 227–35, 237, 241, 245–46, argues that v. 6 is dated after 621 and v. 7 unites vv. 3 and 10, focusing on the theme of destruction.

53. Ibid., pp. 337–43, believes an original autobiographical oracle has been reworked by the Bethel editor after the Josiah reform of 621.

54. On the place of Bethel in the administration of Israel, see G. W. Ahlström, *Royal Administration and National Religion in Ancient Palestine*, pp. 57–61.

55. For information on archaeological discoveries at Bethel, see J. L. Kelso, S.V. "Bethel (Sanctuary)"; also his article, "Bethel."

56. Wolff, *Joel and Amos*, pp. 152–53, 219.

57. Mays, *Amos*, p. 74.

58. For the following outline of 5:21–24, see ibid., pp. 106–8.

59. Wolff, *Joel and Amos*, p. 263.

60. Ibid., p. 232.

61. Kraus, *Worship in Israel*, pp. 154–59.

62. Yohanan Aharoni, *The Land of the Bible: A Historical Geography*, pp. 323–24.

63. For information on archaeological discoveries at Samaria, see G. W. Van Beek, S.V. "Samaria"; J. B. Hennessy, S.V. "Samaria"; N. Avigad, S.V. "Samaria."

64. See p. 91.

65. For information on archaeological discoveries at Dan, see G. W. Van Beek, S.V. "Dan"; A. Biran, S.V. "Dan (city)"; A. Biran, S.V. "Dan, Tel"; Barstad, *Religious Polemics of Amos*, pp. 190–91.

66. Yohanan Aharoni, "The Horned Altar of Beer-sheba."

67. Yigael Yadin, "Beer-sheba: The High Place Destroyed by King Josiah," believes this sacred site was a high place dedicated to pagan gods and destroyed by Josiah, not Hezekiah.

68. For information on archaeological discoveries at Beersheba, see S.

Cohen, S.V. "Beer-sheba"; B. Boyd, S.V. "Beer-sheba"; J. Perrot and R. Gophra, S.V. "Beersheba"; Yohanan Aharoni, S.V. "Beersheba, Tel"; Barstad, *Religious Polemics of Amos*, pp. 198–201.

69. See pp. 93–94.

70. Wolff, *Joel and Amos*, p. 239.

71. On the background for the demand for centralization of worship in Jerusalem by Josiah, see J. W. McKay, *Religion in Judah under the Assyrians: 732–609* B.C., p. 88, n. 41. For a discussion of Deuteronomy 12:14, see Baruch Halpern, "The Centralization Formula in Deuteronomy."

72. On the office of the queen mother and its relationship to the *hieros gamos* of Canaanite culture, see G. W. Ahlström, *Aspects of Syncretism in Israelite Religion*, pp. 57–88.

73. Ernest Nicholson, "The Centralization of the Cult in Deuteronomy," believes the political situation during Hezekiah's reign called for centralization of worship in Jerusalem.

74. Numbers 21:6–9 is an etiological passage that attempts to explain the pagan fertility image in the temple by relating it to God's judgment of the people in the wilderness and to the intercessory role of Moses.

75. John Gray, *I and II Kings, A Commentary*, p. 670.

76. H. H. Rowley, "The Prophet Jeremiah and the Book of Deuteronomy," p. 164.

77. Cripps, *Book of Amos*, pp. 115–16; Harper, *Amos and Hosea*, pp. 9–12; Wolff, *Joel and Amos*, p. 121; Mays, *Amos*, pp. 21–22.

78. Artur Weiser, *The Psalms: A Commentary*, pp. 38–44, places Psalm 50 in a preexilic covenant festival setting.

79. I am indebted to the late G. Henton Davies, Principal of Regent's Park College, Oxford University, for calling my attention to this verse.

80. Francis Brown, S. R. Driver, and Charles A. Briggs, *A Hebrew and English Lexicon of the Old Testament*, in which the meaning of *dāraš* within a cultic context is to seek after or inquire of a deity at a sanctuary; similarly, the meaning of *bōʾ* in a cultic context is to come to or enter a sanctuary. On *dāraš*, see also G. Johannes Botterweck and Helmer Ringgren, eds., *Theological Dictionary of the Old Testament*, pp. 293–307; Watts, *Vision and Prophecy in Amos*, pp. 65–66; Kapelrud, *Central Ideas in Amos*, pp. 36–41.

81. *Dāraš* and *biqqēš* are used synonymously; both are translated "seek" in Psalm 24:6.

82. Mays, *Amos*, p. 87.

83. Wolff's total rejection of a cultic summons to a sanctuary in 5:4b is not persuasive. He argues that the imperative of *dāraš* "nowhere in the Old Testament refers to calling upon Yahweh in his sanctuary." (*Joel and Amos*, p. 238) But Psalm 105 undoubtedly was composed to be recited in a major festival, employing the imperative of *dāraš* in v. 4: "Seek the LORD and his strength, seek his presence continually." (Again *dāraš* and *biqqēš* are used

synonymously.) Wolff also ignores the synonymous usage of *dāraš* and *bō'* in Amos 5:4–5 as a cultic summons to worship.

Chapter 6

1. I am indebted to Keith W. Whitelam, *The Just King: Monarchical Judicial Authority in Ancient Israel*, for the overall structure of the development of the premonarchical and monarchical judicial system in Israel.

2. Robert R. Wilson, "Israel's Judicial System in the Preexilic Period," pp. 234–35. See also idem, "Enforcing the Covenant: The Mechanism of Judicial Authority in Early Israel," pp. 62–63, which cites archaeological evidence for a four-room house that may indicate the existence of nuclear families as the basic social unit.

3. C. H. J. de Geus, *The Tribes of Israel: An Investigation into Some of the Presuppositions of Martin Noth's Amphictyony Hypothesis*, p. 136.

4. Later legislation limited the power of the head of the family by requiring him to take such cases as a son accused of being rebellious (Deut. 21:18–21) or a bride accused of not being a virgin (Deut. 22:13–21) to the city gate for trial by the elders.

5. Wilson, "Enforcing the Covenant," p. 63.

6. The classical scholarly study of the trial proceedings of a clan–town in the premonarchical period is Ludwig Köhler's essay, "Justice in the Gate," in *Hebrew Man*, pp. 127–50.

7. The Abrahamic story is the climax of the Elohistic source, dated ninth century B.C.E.; the Book of Ruth is usually considered postexilic in its present form; the trial of Naboth, to be examined later in this chapter, is part of the deuteronomistic history, compiled during the exilic period. Whereas the historical accuracy of these stories may be questioned, they appear to preserve reliable accounts of Israel's legal customs. It is, however, difficult to date when these judicial procedures were practiced.

8. For a detailed analysis of this trial, see Wilson, "Israel's Judicial System in the Preexilic Period," pp. 238–39.

9. Contrast with Donald A. McKenzie, "The Judge of Israel."

10. Possibly the use of a tree for shade indicates a simple court meeting comparable to the city gate. There are two other references to judgments rendered beneath trees. Saul was encamped beneath the pomegranate tree (I Sam. 14:2) during Jonathan's victory at Michmash. Later in the day, the trial of Jonathan for breaking the fast during the battle was held there. On another occasion, Saul sat at Gibeah under the tamarisk tree (22:6) and conducted the trial of Ahimelech and the priests of Nob. When Saul's body was taken from the wall at Jabesh, it was buried beneath the tamarisk tree (31:13) at Jabesh.

It may be significant that all references use either a person's name or the definite article to indicate a specific tree. This could designate a sacred tree or a specific tree for judging cases. Inasmuch as all these references occur during military campaigns, it is most likely that they refer to a shady place a commander selects to establish his camp. The references all indicate the close relationship that existed in Ancient Israel between military and judicial authority.

11. If G. W. Ahlström, *Royal Administration and National Religion in Ancient Palestine*, p. 22, is correct that sanctuaries were also local centers of administration, then Samuel may have had both sacral and civil responsibilities at these sites.

12. Whitelam, *The Just King*, pp. 51–59.

13. Contrast with Anthony Phillips, *Ancient Israel's Criminal Law: A New Approach to the Decalogue*, p. 22.

14. Tryggve N. D. Mettinger, *King and Messiah: The Civil and Sacral Legitimation of the Israelite Kings*, p. 29, n. 70, argues that in secular cases the conduct of the lot was done by laity while in sacral cases it was done by priests.

15. Although the Urim and Thummim are not mentioned, most critics believe this was the way the selection process was carried out.

16. Roland de Vaux, *Ancient Israel: Its Life and Institutions*, p. 150.

17. See Joseph Blenkinsopp, "Jonathan's Sacrilege. I Sam. 14, 1–46: A Study in Literary History."

18. Whitelam, *The Just King*, p. 74.

19. Following the Septuagint translation not the Masoretic text.

20. See Mettinger, *King and Messiah*, pp. 108–10.

21. Additional evidence of a serious conflict between Saul and Jonathan is Saul's accusation that both his own tribe of Benjamin and his son were involved in a conspiracy to secure the throne of David (I Sam. 22:7–8).

22. This unit appears to be part of a literary conflation of three separate stories: Saul's court at Gibeah (v. 6), charges against the Benjaminites and Jonathan (vv. 7–8), and the trial of Ahimelech and the priests of Nob followed by the flight of Abiathar to David (vv. 9–23).

23. Georg Christian Macholz, "Die Stellung des Königs in der israelitischen Gerichtsverfassung," p. 162.

24. It may be that being priests by royal appointment they were permitted to consult oracles only for officials of the crown.

25. Macholz, "Die Stellung des Königs in der israelitischen Gerichtsverfassung," pp. 160–62.

26. Whitelam, *The Just King*, pp. 123–36.

27. Macholz, "Die Stellung des Königs in der israelitischen Gerichtsverfassung," p. 165, argues that the offense took place in Jerusalem, hence coming under the king's jurisdiction.

28. Contrast with ibid., pp. 165–66.

29. Phillips, *Ancient Israel's Criminal Law*, p. 135, n. 33, argues that the

king was subject to the criminal law but was outside the jurisdiction of the courts. Whitelam, *The Just King*, p. 129, believes that the king was subject only to divine justice.

30. Mettinger, *King and Messiah*, p. 30.

31. Ibid., p. 122.

32. This is also true of the case of the two harlots in I Kings 3:16–28. II Kings 8:3 deals with the restoration of family property that had been formerly included in the crown's estates, hence naturally coming under royal jurisdiction.

33. Norman K. Gottwald, *The Tribes of Yahweh: A Sociology of the Religion of Liberated Israel, 1250–1050 B.C.E.*, p. 369, sees David as a supporter of the local tribal courts.

34. Of course the deuteronomistic historians regarded Absalom's rebellion as sinful. According to Walter Brueggemann, so did the Yahwistic writers who interpreted the Noah story in light of Absalom's rebellion. Brueggemann, in "Kingship and Chaos (A Study in Tenth Century Theology)," argues that the establishment of the orderly world (Gen. 8:33, a royal decree) over against a chaotic flood reflects the Yahwistic writers' interpretation of David's restoration to the throne following the defeat of Absalom.

35. Tomoo Ishida, *The Royal Dynasties in Ancient Israel: A Study on the Formation and Development of Royal-Dynastic Ideology*, pp. 152, 155.

36. Solomon probably served as coregent with David. Solomon's reign begins in I Kings 1:38 while David's death is recorded in 2:10.

37. Mettinger, *King and Messiah*, p. 244.

38. It is also implied in the Book of Judges, because before there were kings in Israel "every man did what was right in his own eyes" (Judg. 17:6, 21:25).

39. Aubrey R. Johnson, *Sacral Kingship in Ancient Israel*, pp. 8–9, develops the relationship between a king's just rule and the economic well-being of the people. See also ibid., "The Role of the King in the Jerusalem Cultus," pp. 76–79.

40. See pp. 39–40. Compare Mettinger, *King and Messiah*, pp. 238–46.

41. David had nothing to fear from Mephibosheth, because this grandson of Saul was disqualified for the throne by his physical disabilities.

42. Mettinger, *King and Messiah*, p. 119, n. 3a, argues that the material on Saul's accession–amnesty policy is redactional. But he admits the practice is presupposed.

43. Although these passages showing the generosity of David and Solomon are part of the pro-Davidic and pro-Solomonic narratives, they nevertheless contain illustrations of the practices of land grants and accession–amnesties that existed at the time.

44. For arguments limiting the narrative to I Kings 21:1–19a see J. Maxwell Miller, "The Fall of the House of Ahab," p. 312.

45. Whitelam, *The Just King*, p. 174.

46. Phillips, *Ancient Israel's Criminal Law*, p. 43.

47. Georgio Buccellati, *Cities and Nations in Ancient Syria*, pp. 131–33.

48. Whitelam, *The Just King*, pp. 181–83.

49. Buccellati, *Cities and Nations of Ancient Syria*, pp. 186–87; Tryggve N.D. Mettinger, *Solomonic State Officials: A Study of the Civil Government Officials of the Israelite Monarchy*, p. 84.

50. de Vaux, *Ancient Israel*, pp. 124–25, and Mettinger, *Solomonic State Officials*, p. 82, argue that abandoned property becomes part of crown lands.

51. Gerhard von Rad, *Deuteronomy, a Commentary*, pp. 40, 114, 117–18; Brevard S. Childs, *The Book of Exodus, a Critical, Theological Commentary*, p. 324; Rolf Knierim, "Exodus 18 und die Neuordnung der mosaischen Gerichtsbarkeit."

52. Peter R. Ackroyd, *I and II Chronicles, Ezra, and Nehemiah*, p. 142.

53. Georg Christian Macholz, "Zur Geschichte der Justizorganisation in Juda," p. 324.

54. Phillips, *Ancient Israel's Criminal Law*, p. 158.

55. Phillips (ibid., p. 23) believes this verse refers to the Persian period.

56. For a brief summary of this position, see Norman K. Gottwald, "Domain Assumption and Societal Models in the Study of Premonarchical Israel."

57. Willy Schottroff, "The Prophet Amos: A Socio-Historical Assessment of His Ministry"; Klaus Koch, *The Prophets: Volume One, the Assyrian Period*, pp. 44–50.

58. Robert B. Coote, *Amos among the Prophets: Composition and Theology*, pp. 24–32.

59. See p. 30.

60. Gottwald, "Domain Assumption and Societal Models," p. 92; Coote, *Amos among the Prophets*, p. 34.

61. On the messianic fulfillment of the promises to David, see Johnson, *Sacral Kingship in Ancient Israel*, pp. 128–44.

62. Many critics believe Micah 5:2–4 [H 5:1–3] is exilic or postexilic. See James L. Mays, *Micah, A Commentary*, pp. 111–17. Delbert R. Hillers, *Micah, a Commentary on the Book of the Prophet Micah*, pp. 64–75, holds that it is an authentic Micah oracle based on traditional material and reworked for an exilic situation.

63. This is especially true if J. Maxwell Miller and John H. Hayes, in *A History of Ancient Israel and Judah*, are correct that Israel and Judah were virtually two separate countries at this time.

64. For evidence supporting the historicity of this system of fortifications, see Jacob M. Myers, *II Chronicles*, pp. 69–70. Ackroyd, *I & II Chronicles, Ezra, Nehemiah*, pp. 128–32, argues that the origin of the list of fortifications cannot be determined, but that the material was used by the chronicler to

illustrate the firm establishment of the kingdom under Rehoboam preceding his faithlessness (II Chron. 12:1–12) that led to God's punishment through Shishak's invasion of the region.

65. Ackroyd, *I & II Chronicles, Ezra, Nehemiah*, pp. 147–52, believes the references to Mennites (II Chron. 20:1; cf. I Chron. 4:41 and II Chron. 26:7) support the position that the account deals with warfare in the Nabataean area of the Negeb in the postexilic period. Another theory that the chronicler had recast II Kings 3 is no longer widely accepted.

66. Ernest W. Nicholson, *The Book of the Prophet Jeremiah, Chapters 1–25*, p. 7, accepts as historically accurate the chronicler's reference to Tekoa as one of the cities Rehoboam fortified and argues that the Jeremiah reference supports the position that it is still an important garrison. John Bright, *Jeremiah*, p. 47, notes that Tekoa was chosen because of its similar sound to the Hebrew imperative *tiqʿū*, "blow."

67. Otto Kaiser, *Isaiah 1–12, a Commentary*, pp. 18–21, 40–42, 44–45.

68. Scholars differ on the date of Micah. Hans Walter Wolff, *Micah the Prophet*, and Hillers, *Micah*, believe his early prophecies date before the fall of Samaria in 721/2. Mays, *Micah*, holds that references to Israel are redactional; he dates the authentic Micah oracles ca. 701. All agree that Micah's bitter attack on leaders and rulers is directed against those in authority in Judah.

69. See Bernhard Lang, "The Social Organization of Peasant Poverty in Biblical Israel," and revised in his *Monotheism and the Prophetic Minority*, pp. 114–27.

70. Koch, *The Prophets: Volume One, the Assyrian Period*, p. 45, maintains that loss of inherited land meant also that the person had no legal status, could not undertake military service or take an active part in cultic life. He cites no supporting evidence.

71. Mays, *Amos*, p. 45.

72. Hans Walter Wolff, *Joel and Amos*, p. 165.

73. Mays, *Amos*, p. 45.

74. Wolff, *Joel and Amos*, p. 166.

75. See pp. 88–91.

76. Wolff, *Joel and Amos*, pp. 324–25.

77. For example, Mays, *Amos*, pp. 142–46; Erling Hammershaimb, *The Book of Amos, a Commentary*, pp. 121–25; Richard S. Cripps, *A Critical & Exegetical Commentary on the Book of Amos*, pp. 242–46; William Rainey Harper, *A Critical and Exegetical Commentary on Amos and Hosea*, pp. 176–79.

78. Arvid S. Kapelrud, "New Ideas in Amos," pp. 200–205. It is usually held that the identification of poverty with righteousness and wealth with sinfulness is an idea that arose in the postexilic period when cooperation with foreign powers was necessary to achieve political and economic prominence.

79. For example, Coote, *Amos among the Prophets*, p. 34; Gustavo Gutiérrez, *A Theology of Liberation: History, Politics, and Salvation*, pp. 291–96; Elsa Tamez, *Bible of the Oppressed*; Willy Schottroff, "The Prophet Amos: A Socio-Historical Assessment of His Ministry," pp. 27–46.

80. The Hebrew means either "great houses" or "many houses." In modern society, the practice among the affluent of owning more than one home forms a striking parallel to what Amos condemned.

81. Wolff, *Joel and Amos*, p. 247. The first building in Ancient Israel constructed of cut stone was Solomon's temple. The practice was later employed in constructing homes for the wealthy.

82. See p. 103.

83. This may refer to a *marzēaḥ* festival. See above pp. 88–91.

84. Contrast the characterization of a good woman in Proverbs 31:10–31.

85. For this interpretation of 2:7b, see Wolff, *Joel and Amos*, pp. 166–67. For its possible connection with a *marzēaḥ* festival, see above, pp. 89–90.

86. The reference to wine acquired by fining someone and subsequently drunk at temple festivals is apparently a similar offense to God (2:8b). The exact parallel to garments taken in pledge is not known. Again, it may refer to a *marzēaḥ* festival.

87. Wormwood is a small bushy plant that is not poisonous but is extremely bitter to the taste.

88. Following Mays' organization in *Amos*, pp. 91–92. The first two texts are indictments and the second two are exhortations.

89. E. R. Achtemeier, "Righteousness in the Old Testament."

90. See Mays, *Amos*, pp. 90–95, for this identification of the woe oracle. Against this see Wolff, *Joel and Amos*, p. 233.

91. Wolff, *Joel and Amos*, p. 245.

92. Norman K. Gottwald, *The Hebrew Bible—A Socio-Literary Introduction*, pp. 357–58.

93. Hans Walter Wolff, *Hosea, A Commentary on the Book of the Prophet Hosea*, p. 97.

94. Ahlström, *Royal Administration and National Religion in Ancient Palestine*, pp. 51–65.

95. It may be significant that the word used for the estates owned by Abiathar (I Kings 2:20–27) and Jeremiah (Jer. 32:6–15) is "field" (*śādeh*), while the word used for Amaziah's estate is "land" (*'ădāmâ*). The Amaziah reference probably designates a royal land grant and not inherited land.

Chapter 7

1. Hans Walter Wolff, "Das Thema 'Umkehr' in der alttestamentlichen Prophetie."

2. Hans Walter Wolff, "Das Kerygma des deuteronomistischen Geschichtswerks," English trans.: "The Kerygma of the Deuteronomistic Historical Work."

3. See A. Vanlier Hunter, *Seek the Lord! A Study of the Meaning and Function of the Exhortation in Amos, Hosea, Isaiah, Micah, and Zephaniah*, for a detailed critique of "the majority opinion" and well-documented support of Wolff's position.

4. Abraham J. Heschel, *The Prophets*, p. 12.

5. Ibid., p. 12, n. 5.

6. C. F. Whitley, *The Prophetic Achievement*, p. 155.

7. For an excellent discussion of the deuteronomistic interpretation of prophecy, see Hunter, *Seek the Lord!*, pp. 39–55.

8. Ibid., pp. 51–53.

9. Wolff, "Kerygma of the Deuteronomistic Historical Work."

10. For a brief survey of various theories on the repentance motif in Amos, see Klaus Koch, *The Prophets: Volume One, the Assyrian Period*, pp. 41–44.

11. John Marsh, *Amos and Micah: Thus Saith the Lord*, p. 51. For a similar position, see Jacob Myers, *Hosea, Joel, Amos, Obadiah, Jonah*, p. 122; Heschel, *Prophets*, pp. 36–37.

12. William Rainey Harper, *A Critical and Exegetical Commentary on Amos and Hosea*, p. 104.

13. Walter Brueggemann, "Amos IV 4–13 and Israel's Covenant Worship." See also Myers, *Hosea, Joel, Amos, Obadiah, Jonah*, p. 122.

14. See also James L. Crenshaw, "Amos and the Theophanic Tradition," pp. 212–13. Crenshaw accepts 4:6–11 as a genuine Amos oracle that recites the judgmental actions of God, which failed to elicit Israel's repentance. He believes the passage contains theophanic overtones related to holy war.

15. Hans Walter Wolff, *Joel and Amos*, pp. 211–25.

16. George Ramsey, "Amos 4:12—A New Perspective."

17. Hunter, *Seek the Lord!*, p. 121.

18. To take the three doxologies (4:13; 5:8–9; 9:5–6) as postexilic additions does not alter the judgmental nature of the oracle.

19. Thomas M. Raitt, "The Prophetic Summons to Repentance" (Vanderbilt University, Ph.D. dissertation). A. Vanlier Hunter, "Seek the Lord! A Study of the Meaning and Function of the Exhortation in Amos, Hosea, Isaiah, Micah, and Zephaniah" (doctoral dissertation to the Theological Faculty of the University of Basel in 1981).

20. Hunter, *Seek the Lord!*, pp. 102–5.

21. Wolff, *Joel and Amos*, p. 232.

22. Wilhelm Rudolph, *Joel–Amos–Obadiah–Jona*, p. 184.

23. Thomas M. Raitt, "The Prophetic Summons to Repentance."

24. Ibid., pp. 31–32.

25. Wolff, *Joel and Amos*, pp. 233, 235, 240–41.

26. Ibid., pp. 234, 235, 250–51.

27. See William L. Holladay, *The Root Šûbh in the Old Testament*, pp. 128–39.

28. Hunter, *Seek the Lord!*, pp. 61–67.

29. Curt Kuhl, in *The Prophets of Israel*, writes: "Moreover, we find in the prophetic message the admonition-oracle (e.g. Am. v. 5–6) the object of which is to persuade its hearers to return to Yahweh and His obedience, in the hope that punishment will be turned aside" (p. 31).

30. Walter Eichrodt, in *Theology of the Old Testament*, writes: "The eschatological hope of salvation does nothing to limit the seriousness of the judgment; on the contrary, it is what gives it its full severity. For this hope looks for a genuine new creation by Yahweh after the old order has been totally destroyed" (p. 379, n. 2).

31. Gerhard von Rad, *The Problem of the Hexateuch and Other Essays*, p. 254.

32. On the basis of Genesis 45:7, Hunter, in *Seek the Lord!*, pp. 90–94, argues that the *šĕʾērît* in Amos means "descendants" not "remnant."

33. Because the two dominant tribes in the northern kingdom were Manasseh and Ephraim (in patriarchal narratives the sons of Joseph), the term Joseph was used by Amos to designate Israel.

34. Hunter, *Seek the Lord!*, p. 70.

35. James L. Mays, *Amos, a Commentary*, pp. 87–88.

36. Arvid S. Kapelrud, *Central Ideas in Amos*, p. 36.

37. Ibid., pp. 25, 27, 42–47.

38. For a critique of this position, see Hans Gottlieb, "Amos und Jerusalem," esp. pp. 451–54, and Erling Hammershaimb, *The Book of Amos, a Commentary*, pp. 77–78.

39. Wolff, *Joel and Amos*, pp. 260–62, 266.

40. Hunter, *Seek the Lord!*, p. 114.

41. Wolff, in *Joel and Amos*, p. 262, suggests the more provocative translation "savor" instead of the more common "take pleasure in."

42. See Mays, *Amos*, p. 110, and Hunter, *Seek the Lord!*, p. 108, for a defense of separating the burnt offerings from the gift offerings, considering the latter alone as part of the clause that follows.

43. Robert B. Coote, *Amos among the Prophets*, pp. 56–57. For a similar arrangement, see Hammershaimb, *Book of Amos*, pp. 107–8.

44. Ernst Würthwein, "Amos-Studien."

45. Mays, *Amos*, p. 128.

46. John D.W. Watts, *Vision and Prophecy in Amos*, pp. 36–45.

47. Coote, *Amos among the Prophets*, pp. 92–93.

48. Walter Brueggemann, "Amos' Intercessory Formula."

49. I agree with Robert P. Carroll, *When Prophecy Failed*, who writes

concerning the prophetic view of repentance: "I suspect that it may have been a prophetic invention put forward as a last ditch possibility for a corrupt society and one which took until the deuteronomists to become a formal principle" (p. 24). James M. Ward, *Amos & Isaiah: Prophet of the Word of God*, pp. 47–48 argues that although Amos called for repentance, he did not believe destruction could be avoided. Amos believed, however, that destruction might be more than a catastrophe if it occasioned moral renewal. For the purifying nature of repentance, see Hammershaimb, *Book of Amos*, pp. 136–38.

Chapter 8

1. The publications on the Day of Yahweh are extensive. See A. Joseph Everson, "The Days of Yahweh," and idem, "Day of the Lord," which includes a bibliography; C. Van Leeuwen, "The Prophecy of the Yom YHWH in Amos V 18–20."

2. Sigmund Mowinckel, *He That Cometh*, p. 145.

3. Gerhard von Rad, *Old Testament Theology, Vol. II, The Theology of Israel's Prophetic Traditions*, p. 123.

4. Frank M. Cross, Jr., "The Divine Warrior in Israel's Early Cult," p. 16. Cross indicates that the earliest sources do not emphasize the sea crossing. The role of the sea was eventually stressed because it could be related to the cosmogonic battle between God and the sea monster.

5. Ibid., p. 30.

6. Hans Walter Wolff, *Joel and Amos*, pp. 175, 324, 329–30 believes 8:9 is redactional. He considers "and on that day it will come to pass" (*wĕhāyâ bayyôm hahû'*) stylistically different. Some scholars also argue that vv. 9–10 contain postexilic apocalyptic imagery. Such imagery, however, is part of the holy-war and divine-warrior motif of the preexilic period. I consider these verses authentic Amos.

7. "The days are coming" also appears in 8:11–12 and 9:13–15. Both these units are usually regarded as redactional. Amos 8:11–12 reveals the hand of the deuteronomists and 9:13–15 is a postexilic hope addition. See Wolff, *Joel and Amos*, pp. 324–26, 330–31, 350–55.

8. Delbert R. Hillers, "Amos 7,4 and Ancient Parallels," p. 222.

9. James B. Pritchard, ed., *The Ancient Near East in Pictures Relating to the Old Testament*, figure 490.

10. Patrick D. Miller, Jr., "Fire in the Mythology of Canaan and Israel," pp. 259–60.

11. The Hebrew word *yāṣā'*, "to go forth," frequently refers to going into battle.

12. On loss of land as a divine punishment, see Walter Brueggemann, *The Land: Place as Gift, Promise, and Challenge in Biblical Faith*.

13. Variant forms of the divine name "God of Hosts" appear in the book of Amos (3:13; 4:13; 5:8 [in LXX], 14, 15, 16, 27; 6:8, 14; 9:5 [shortened version]). If "Hosts" refers to Israel's armies or to the starry hosts, it would have military connotations. Wolff, however, in *Joel and Amos*, pp. 287–88, maintains all these references are from the hand of redactors.

14. The hiphil form of *nākâ*, "to smite, strike, or wreck," is especially appropriate for a natural catastrophe such as an earthquake.

15. Following Wolff's interpretation in *Joel and Amos*, pp. 282–83.

16. Robert B. Coote, *Amos among the Prophets*, indicates that in his A stage of the Book of Amos only the ruling elite will be annihilated. Presumably the poor will remain to occupy the land.

Works Cited

Articles

Aberbach, Moses and Leivy Smolar. "Aaron, Jeroboam, and the Golden Calves." *JBL* 86 (1967): 129–40.

Achtemeier, E. R. "Righteousness in the Old Testament." *IDB*. Gen. ed. George Arthur Buttrick. Nashville, Tenn.: Abingdon Press, 1962. Vol. 4, pp. 80–85.

Ackroyd, Peter R. "The Meaning of Hebrew *dwr* Reconsidered." *JSS* 13 (1968): 3–10.

Aharoni, Yohanan. "Beersheba, Tel." *Encyclopedia of Archaeological Excavations in the Holy Land*. Eds. Michael Avi-Yonah and Ephraim Stern. Englewood Cliffs, N.J.: Prentice-Hall, Inc., 1975. Vol. I, pp. 160–68.

Aharoni, Yohanan. "The Horned Altar of Beer-sheba." *BA* 37 (1974): 2–6.

Anderson, Bernhard W. "Host, Host of Heaven." *IDB*. Gen. ed. George Arthur Buttrick. Nashville, Tenn.: Abingdon Press, 1962. Vol. 2, pp. 654–56.

Attridge, Harold W. and Robert A. Oden, Jr. "Philo of Byblos: The Phoenician History." Monograph series 9. *CBQ* (1981).

Avigad, N. "The Priest of Dor." *IEJ* 25 (1975): 101–5.

Avigad, N. "Samaria." *Encyclopedia of Archaeological Excavations in the Holy Land*. Eds, Michael Avi-Yonah and Ephraim Stern. Englewood Cliffs, N.J.: Prentice-Hall, Inc., 1975. Vol. IV, pp. 1032–50.

Bartlett, J. R. "Zadok and His Successors at Jerusalem." *JTS*. New series, 19 (1968): 1–18.

Ben-Barak, Zafrira. "The Legal Background to the Restoration of Michal

213

to David." *Studies in the Historical Books of the Old Testament*. Ed. J. A. Emerton. VTSup 20. Leiden, The Netherlands: E. J. Brill, 1979. pp. 15–29.

Bentzen, Aage. "The Ritual Background of Amos i 2–ii 6." *OS* 8 (1950): 85–99.

Biran, A. "Dan (city)." *IDB*. Gen. ed. Keith Crim. Nashville, Tenn.: Abingdon Press, 1976. Suppl. vol., p. 205.

Biran, A. "Dan, Tel." *Encyclopedia of Archaeological Excavations in the Holy Land*. Eds. Michael Avi-Yonah and Ephraim Stern. Englewood Cliffs, N.J.: Prentice-Hall, Inc., 1975. Vol. I, pp. 313–21.

Blenkinsopp, Joseph. "Jonathan's Sacrilege. I Sam. 14, 1–46: A Study in Literary History." *CBQ* 2 (1964): 423–49.

Boyd, B. "Beer-sheba," *IDB*. Ben. ed. Keith Crim. Nashville, Tenn.: Abingdon Press, 1976. Suppl. vol., pp. 93–95.

Brueggemann, Walter. "Amos IV 4–13 and Israel's Covenant Worship." *VT* 15 (1965): 1–15.

Brueggemann, Walter. "Amos' Intercessary Formula." *VT* 19 (1969): 385–99.

Brueggemann, Walter. "Kingship and Chaos (A Study in Tenth Century Theology)." *CBQ* 33 (1971): 317–32.

Clifford, Richard J. "The Tent of El and the Israelite Tent of Meeting." *CBQ* 33 (1971): 221–27.

Cohen, S. "Beer-sheba." *IBD*. Gen. ed. George Arthur Buttrick. Nashville, Tenn.: Abingdon Press, 1962. Vol. 1, pp. 375–76.

Cook, G. "The Sons of (the) God(s)." *ZAW* 76 (1964): 22–47.

Coote, Robert B. "Amos 1:11: RHMYW." *JBL* 90 (1971): 206–8.

Craigie, Peter C. "Amos the *nōqēd* in the Light of Ugarit." *SR* 11 (1982): 29–33.

Crenshaw, James L. "Amos and the Theophanic Tradition." *ZAW* 80 (1968): 203–15.

Crenshaw, James L. "The Influence of the Wise upon Amos." *ZAW* 79 (1967): 42–52.

Cross, Frank M., Jr. "The Council of Yahweh in Second Isaiah." *JNES* 12 (1953): 274–77.

Cross, Frank M., Jr. "The Divine Warrior in Israel's Early Cult." *Biblical Motifs: Origins and Trans-formations*. Ed. Alexander Altmann. Cambridge, Mass.: Harvard University Press, 1966. Pp. 11–30.

Cross, Frank M., Jr. and David N. Freedman. "The Blessing of Moses." *JBL* 67 (1948): 191–210.

Crusemann, Frank. "Kritik an Amos im deuteronomistischen Geschichtswerk: Erwägungen zu 2. Könige 14:27." *Probleme Biblischer Theologie*. Festschrift für Gerhard von Rad. Ed. Hans Walter Wolff. München: Chr. Kaiser Verlag, 1971. Pp. 57–63.

Davies, G. Henton. "Amos—the Prophet of Re-Union." *Exp Tim* 92 (1981): 196–200.

Davies, G. Henton. "Ark of the Covenant." *IDB*. Gen. ed. George Arthur Buttrick. Nashville, Tenn.: Abingdon Press, 1962. Vol. l, pp. 222–26.

Driver, S. R. "Lord of Hosts." *Hasting's Dictionary of the Bible*. New York: Charles Scribner's Sons, 1900. Vol. 3, pp. 137–38.

Eakin, Frank E., Jr. "Yahwism and Baalism before the Exile." *JBL* 84 (1965): 407–14.

Everson, A. Joseph, "Day of the Lord." *IDB*. Gen. ed. Keith Crim. Nashville, Tenn.: Abingdon Press, 1976. Suppl. vol., pp. 209–10.

Everson, A. Joseph. "The Days of Yahweh." *JBL* 93 (1974): 329–37.

Fensham, F. C. "The treaty between the Israelites and Tyrians." VTSup 17. Leiden, The Netherlands: E. J. Brill, 1969. Pp. 71–87.

Fishbane, Michael. "Additional Remarks on RHMYW (Amos 1:11)." *JBL* 91 (1972): 391–93.

Fishbane, Michael. "The Treaty Background of Amos 1:11 and Related Matters." *JBL* 89 (1970): 313–18.

Flanagan, James W. "Chiefs in Israel." *JSOT* 20 (1981): 47–73.

Gordis, Robert. "The Composition and Structure of Amos." *HTR* 33 (1940): 239–51.

Gottlieb, Hans. "Amos und Jerusalem." *VT* 17 (1967): 430–63.

Gottwald, Norman K. "Domain Assumption and Societal Models in the Study of Premonarchical Israel." VTSup 28. Leiden, The Netherlands: E. J. Brill, 1975. Pp. 89–100.

Gottwald, Norman K. "Two Models for the Origins of Ancient Israel: Social Revolution or Frontier Development." *The Quest for the Kingdom of God: Studies in Honor of George E. Mendenhall*. Eds. H. B. Huffmon, F. A. Spina, and A. R. W. Green. Winona Lake, Ind.: Eisenbauns, 1983. Pp. 5–24.

Halpern, Baruch. "The Centralization Formula in Deuteronomy." *VT* 31 (1981): 20–38.

Haran, Menahem. "Observations on the Historical Background of Amos 1:2–2:6." *IEJ* 18 (1968): 201–12.

Haran, Menahem. "The Rise and Decline of the Empire of Jeroboam ben Joash." *VT* 17 (1967): 266–97.

Hayes, John H. "The Tradition of Zion's Inviolability," *JBL* 82 (1963): 419–26.

Hayes, John H. "The Usage of Oracles against Foreign Nations in Ancient Israel." *JBL* 87 (1968): 81–92.

Hennessy, J. B. "Samaria." *IDB*. Gen. ed. Keith Crim. Nashville, Tenn.: Abingdon Press, 1976. Suppl. vol., pp. 771–72.

Hillers, Delbert R. "Amos 7,4 and Ancient Parallels." *CBQ* 26 (1964): 221–25.

Hobbs, T. R. "Amos 3,1b and 2, 10." *ZAW* 81 (1969): 384–87.

Hollis, F. J. "The Sun Cult and the Temple at Jerusalem." *Myth and Ritual.* Ed. Samuel Henry Hooke. Oxford: Oxford University Press, 1937. Pp. 87–110.

Jacobsen, Thorkild. "Primitive Democracy in Mesopotamia." *JNES* 2 (1943): 159–72.

Johnson, Aubrey R. "Hebrew Conceptions of Kingship." *Myth, Ritual, and Kingship.* Ed. Samuel Henry Hooke. Oxford: Clarendon Press, 1958. Pp. 204–35.

Johnson, Aubrey R. "The Role of the King in the Jerusalem Cultus." *The Labyrinth.* Ed. Samuel Henry Hooke. New York: Macmillan Company, 1935. Pp. 73–111.

Kapelrud, Arvid S. "New Ideas in Amos." VTSup 15. Leiden, The Netherlands: E. J. Brill, 1966. Pp. 193–206.

Kapelrud, Arvid S. "Temple Building, a Task for Gods and Kings," *Or* 32 (1963): 56–62.

Kapelrud, Arvid S. "Ugarit." *IDB.* Gen. ed. George Arthur Buttrick. Nashville, Tenn.: Abingdon Press, 1962. Vol. 4, pp. 724–32.

Kelso, J. L. "Bethel." *Encyclopedia of Archaeological Excavations in the Holy Land.* Eds. Michael Avi-Yonah and Ephraim Stern. Englewood Cliffs, N.J.: Prentice-Hall, Inc., 1975. Vol. I, pp. 190–93.

Kelso, J. L. "Bethel (Sanctuary)." *IDB.* Gen. ed. George Arthur Buttrick. Nashville, Tenn.: Abingdon Press, 1962. Vol. 1, pp. 391–93.

Knierim, Rolf. "Exodus 18 und die Neuordnung der mosaischen Gerichtsbarkeit." *ZAW* 73 (1961): 146–71.

Lang, Bernhard. "The Social Organization of Peasant Poverty in Biblical Israel." *JSOT* 24 (1982): 47–63.

Lindars, Barnabas. "The Israelite Tribes in Judah." *Studies in the Historical Books of the Old Testament.* Ed. J. A. Emerton. VTSup 30. Leiden, The Netherlands: E. J. Brill, 1979. Pp. 95–112.

Lorton, David. "Toward a Constitutional Approach to Ancient Egyptian Kingship." *JAOS* 99 (1979): 460–65.

Macholz, Georg Christian. "Die Stellung des Königs in der Israelitischen Gerichtsverfassung." *ZAW* 84 (1972): 157–82.

Macholz, Georg Christian. "Zur Geschichte der Justizorgansation in Juda." *ZAW* 84 (1972): 314–40.

Mauchline, John. "Implicit Signs of a Persistent Belief in the Davidic Empire." *VT* 20 (1970): 287–303.

McKenzie, Donald A. "The Judges of Israel." *VT* 17 (1967): 118–21.

Mendenhall, George. "The Hebrew Conquest of Palestine." *BA* 25 (1962): 66–87.

Mendenhall, George. "Social Organization in Early Israel." *Magnalia Dei: The Mighty Acts of God.* Eds. Frank M. Cross, Jr., Werner E. Lemke, and Patrick D. Miller, Jr. Garden City, N.Y.: Doubleday & Company, Inc., 1976. Pp. 132–51.

Mettinger, Tryggve N. D. 'YHWH SABAOTH—the Heavenly King on the Cherubim Throne." *Studies in the Period of David and Solomon and Other Essays*. Ed. Tomoo Ishida. Winona Lake, Ind.: Eisenbauns, 1982. Pp. 109–38.

Milgrom, Jacob. "Religious Conversion and the Revolt Model for the Formation of Israel." *JBL* 101 (1982): 169–76.

Miller, J. Maxvell. "The Elisha Cycle and the Accounts of the Omride Wars." *JBL* 85 (1966): 441–54.

Miller, J. Maxwell. "The Fall of the House of Ahab." *VT* 17 (1967): 307–24.

Miller, J. Maxwell. "The Israelite Occuption of Canaan." *Israelite and Judaean History*. Eds. John H. Hayes and J. Maxwell Miller. Philadelpia: The Westminster Press, 1977. Pp. 213–84.

Miller, Patrick D., Jr. "El the Warrior." *HTR* 60 (1967): 411–31.

Miller, Patrick D., Jr. "Fire in the Mythology of Canaan and Israel." *CBQ* 27 (1965): 256–61.

Morgenstern, Julian. "Three Calendars of Ancient Israel." *HUCA* I (1924): 67–71.

Mowinckel, Sigmund. "The Spirit and the Word in the Pre-exilic Reforming Prophets." *JBL* 53 (1934): 199–227.

Muilenburg, James. "Introduction and Exegesis of Isaiah 40–66." *IB*. Gen. ed. George Arthur Buttrick. Nashville, Tenn.: Abingdon Press, 1956. Vol. 5, pp. 422–34.

Neuberg, Frank J. "An Unrecognized Meaning of Hebrew DOR." *JNES* 9 (1950): 215–17.

Nicholson, Ernest. "The Centralization of the Cult in Deuteronomy." *VT* 13 (1963): 380–89.

North, C. R. "Pentateuchal Criticism." *The Old Testament and Modern Study: A Generation of Discovery and Research*. Ed. H. H. Rowley. Oxford: Clarendon Press, 1951. Pp. 48–83.

Obermann, J. "The Divine Name YHWH in the Light of Recent Discoveries." *JBL* 68 (1949): 309–14.

Oliver, J. P. J. "In Search of a Capital for the Northern Kingdom." *JNSL* 11 (1983): 117–32.

Paul, Shalom M. "Sargon's Administrative Diction in II Kings 17, 27." *JBL* 88 (1969): 73–74.

Perrot, J. and R. Gophra. "Beersheba." *Encyclopedia of Archaeological Excavations in the Holy Land*. Eds. Michael Avi-Yonah and Ephraim Stern. Englewood Cliffs, N.J.: Prentice-Hall, Inc., 1975. Vol. I, pp. 153–59.

Polley, Max E. "The Call and Commission of the Hebrew Prophets in the

Council of Yahweh, Examined in Its Ancient Near Eastern Setting." *Scripture in Context: Essays on the Comparative Method.* Pittsburgh Theological Seminary Monograph Series. Eds. Carl D. Evans, William W. Hallo, and John B. White. Pittsburgh: The Pickwick Press, 1980. Pp. 141–56.

Pope, Marvin H. "A Divine Banquet at Ugarit." *The Use of the Old Testament in the New and Other Essays.* Ed. James M. Efird. Durham, N.C.: Duke University Press, 1972. Pp. 170–203.

Porter, J. R. "The Interpretation of 2 Samuel VI and Psalm CXXXII." *JTS.* New series, 5 (1954): 161–73.

Priest, John. "The Covenant of Brothers." *JBL* 84 (1965): 400–406.

Rabe, Virgil W. "Israelite Opposition to the Temple." *CBQ* 29 (1967): 228–33.

Raitt, Thomas M. "The Prophetic Summons to Repentance." *ZAW* 83 (1971): 30–49.

Ramsey, George. "Amos 4:12—a New Perspective." *JBL* 89 (1970): 187–91.

Richardson, H. Neil. "SKT (Amos 9:11): 'Booth' or 'Succoth'?" *JBL* 92 (1973): 375–81.

Roberts, J. J. M. "The Davidic Origin of the Zion Tradition." *JBL* 92 (1973): 329–44.

Roberts, J. J. M. "El." *IDB.* Gen. ed. Keith Crim. Nashville, Tenn.: Abingdon Press. 1976. Suppl. vol., pp. 255–58.

Roberts, J. J. M. "Zion in the Theology of the Davidic–Solomonic Empire." *Studies in the Period of David and Solomon and Other Essays.* Ed. Tomoo Ishida. Winona Lake, Ind.: Eisenbauns, 1982. Pp. 93–108.

Roberts, J. J. M. "Zion Tradition." *IDB.* Ed. Keith Crim. Nashville, Tenn.: Abingdon Press, 1976. Suppl. vol., pp. 985–87.

Robinson, H. Wheeler. "The Council of Yahweh." *JTS* 45 (1944): 151–57.

Robinson, H. Wheeler. "Prophetic Symbolism." *Old Testament Essays.* Ed. D. C. Simpson. London: Charles Griffin and Company, Ltd., 1927. Pp. 1–17.

Rosenberg, R. A. "Shalem (God)." *IDB.* Gen. ed. Keith Crim. Nashville, Tenn.: Abingdon Press, 1976. Suppl. vol., pp. 820–21.

Ross, J. P. "Yahweh Ṣĕbāʾôt in Samuel and Psalms." *VT* 17 (1967): 76–92.

Rowley, H. H. "The Prophet Jeremiah and the Book of Deuteronomy." *Studies in Old Testament Prophecy.* Ed. H. H. Rowley. Edinburgh: T. & T. Clark, 1950. Pp. 157–74.

Rowley, H. H. "The Samaritan Schism in Legend and History." *Israel's Prophetic Heritage.* Eds. Bernhard W. Anderson and Walter Harrelson. London: SCM Press, Ltd., 1962. Pp. 208–22.

Rowley, H. H. "The Unity of the Old Testament." *BJRL* 29 (1945–1946): 326–58.

Rowley, H. H. "Was Amos a Nabi?" *Festschrift Otto Eissfeldt*. Ed. Johann Fück. Halle an der Saales, Tübingen: Max Niemeyer Verlag, 1947. Pp. 191–98.

Rowley, H. H. "Zadok and Nehushtan." *JBL* 58 (1939): 113–41.

Schmidt, W. H. "Die deuteronomistische Redaktion des Amosbuches." *ZAW* 77 (1965): 168–92.

Schottroff, Willy. "The Prophet Amos: A Socio-Historical Assessment of His Ministry." *God of the Lowly: Socio-Historical Interpretations of the Bible*. Eds. Willy Schottroff and Wolfgang Stegemann. Trans. Matthew J. O'Connell. Maryknoll, N.Y.: Orbis Books, 1984. Pp. 33–40.

Schoville, Keith N. "A note on the oracles of Amos against Gaza, Tyre, and Edom." *Studies in Prophecy*. VTSup 26. Leiden, The Netherlands: E. J. Brill, 1974. Pp. 55–63.

Shanks, Hershel. "Ancient Ivory: The Story of Wealth, Decadence, and Beauty." *BAR* 11, no. 5 (1985): 40–53.

Stager, Laurence E. "The Archaeology of the Family in Ancient Israel." *BASOR* 260 (1985): 5–11.

Talmon, S. "Divergences in Calendar-Reckoning in Ephraim and Judah." *VT* 8 (1958): 48–74.

Van Beek, G. W. "Dan." *IDB*. Gen. ed. George Arthur Buttrick. Nashville, Tenn.: Abingdon Press, 1962. Vol. 1, pp. 758–60.

Van Beek, G. W. "Samaria." *IDB*. Gen. Ed. George Arthur Buttrick. Nashville, Tenn.: Abingdon Press, 1962. Vol. 4, pp. 182–88.

Van Leeuwen, C. "The Prophecy of the Yom YHWH in Amos V 18–20." *OS* 19 (1974): 113–34.

Vawter, Bruce. "Prophecy and the Redactional Question." *No Famine in the Land: Studies in Honor of John L. McKenzie*. Eds. James W. Flanagan and Anita Weisbrod Robinson. Missoula, Mont.: Scholars Press, 1975. Pp. 127–39.

Weinfeld, Moshe. "Covenant, Davidic." *IDB*. Ed. Keith Crim. Nashville, Tenn.: Abingdon Press, 1976. Suppl. vol., pp. 188–92.

Weinfeld, Moshe. "The Covenant of Grant in the OT and in the Ancient Near East." *JAOS* 90 (1970): 184–203.

Weinfeld, Moshe. "Zion and Jerusalem a Religious and Politcal Capital: Ideology and Utopia." *The Poet and the Historian: Essays in Literary and Historical Biblical Criticism*. Harvard Semitic Studies. Ed. Richard Elliott Friedman. Chico, Calif.: Scholars Press, 1983. Pp. 75–115.

Weiss, M. "The Pattern of the 'Execration Tests' in the Prophetic Literature." *IEJ* 19 (1969): 150–57.

Wilson, Robert R. "Enforcing the Covenant: The Mechanism of Judicial Authority in Early Israel." *The Quest for the Kingdom of God: Studies in Honor of George E. Mendenhall.* Eds. H. B. Huffmon et al. Winona Lake, Ind.: Eisenbrauns, 1983. Pp. 59–75.

Wilson, Robert R. "Israel's Judicial System in the Preexilic Period." *JQR* 74 (1983): 229–48.

Wolff, Hans Walter. "Das Kerygma des deuteronomistischen Geschichtswerks." *ZAW* 73 (1961): 171–86.

Wolff, Hans Walter. "The Kerygma of the Deuteronomistic Historical Work." *The Vitality of Old Testament Tradition.* Trans. Frederick C. Prussner. Eds. Walter Brueggemann and Hans Walter Wolff. Atlanta, Ga.: John Knox Press, 1975. Pp. 83–100.

Wolff, Hans Walter. "Das Thema 'Umkehr' in der alttestamentlichen Prophetie." *ZTK* 48 (1951): 129–48.

Würthwein, Ernst. "Amos-Studien." *ZAW* 62 (1950): 10–52.

Yadin, Yigael. "Beer-sheba: The High Place Destroyed by King Josiah." *BASOR* 222 (1976): 5–17.

Yevin, S. "Did the Kingdom of Israel have a Maritime Policy?" *JQR* 50 (1959–1960): 193–228.

Zevit, Ziony. "Clio, I presume." *BASOR* 260 (1985): 71–82.

Zevit, Ziony. "Deuteronomistic Historiography in 1 Kings 12–2 Kings 17 and the Reinvestiture of the Israelian Cult." *JSOT* 32 (1985): 57–73.

Books

Achtemeier, Paul J. *The Inspiration of Scripture: Problems and Proposals.* Biblical Perspectives on Current Issues. Philadelphia: The Westminster Press, 1980.

Ackroyd, Peter R. *I and II Chronicles, Ezra, and Nehemiah.* Torch Commentary. London: SCM Press, Ltd., 1973.

Aharoni, Yohanan. *The Land of the Bible: A Historical Geography.* Trans. A. F. Rainey. Philadelphia: The Westminster Press, 1967.

Ahlström, G. W. *Aspects of Syncretism in Israelite Religion.* Lund, Sweden: C. W. K. Gleerup, n.d.

Ahlström, G. W. *Royal Administration and National Religion in Ancient Palestine.* Studies in the History of the Ancient Near East, vol. I. Ed. M. H. E. Weippert. Leiden, The Netherlands: E. J. Brill, 1982.

Albright, William F. *From the Stone Age to Christianity.* Baltimore: The Johns Hopkins Press, 1940.

Albright, William F. *Yahweh and the Gods of Canaan: a Historical Analysis of Two Contrasting Faiths.* London: Athlone, 1968.

Alt, Albrecht. *Essays on Old Testament History and Religion.* Trans. R. A. Wilson. Garden City, N.Y.: Doubleday & Company, Inc., 1968. (Originally published in German 1925–1934.)

Anderson, Bernhard W. *Understanding the Old Testament.* 4th ed. Englewood Cliffs, N.J.: Prentice-Hall, Inc., 1986.

Barstad, Hans M. *The Religious Polemics of Amos.* VTSup 34. Leiden, The Netherlands: E. J. Brill, 1984.

Barton, John. *Amos's Oracles against the Nations.* The Society for Old Testament Study Monograph Series, 6. Cambridge: Cambridge University Press, 1980.

Bewer, Julius A. *The Prophets.* New York: Harper & Brothers, 1949.

Beyerlin, Walter, ed. *Near Eastern Religious Texts Relating to the Old Testament.* Philadelphia: The Westminster Press, 1978.

Blenkinsopp, Joseph. *A History of Prophecy in Israel.* Philadelphia: The Westminster Press, 1983.

Botterweck, G. Johannes and Helmer Ringgren, eds. *Theological Dictionary of the Old Testament.* Trans. John T. Willis, Geoffrey W. Bromiley, and David E. Greene. Grand Rapids, Mich.: William B. Eerdmans, 1978. Vol. III.

Bright, John. *A History of Israel.* 3rd ed. Philadelphia: The Westminster Press, 1981.

Bright, John. *Jeremiah.* The Anchor Bible. New York: Doubleday & Company, Inc., 1965.

Brown, Francis, S. R. Driver, and Charles A. Briggs. *A Hebrew and English Lexicon of the Old Testament.* Oxford: Clarendon Press, 1972. (First published in 1907.)

Brueggemann, Walter. *The Land: Place as Gift, Promise, and Challenge in Biblical Faith.* Overtures to Biblical Theology Series. Philadelphia: Fortress Press, 1977.

Buccellati, Georgio. *Cities and Nations of Ancient Syria.* Studi Semitic 26. Rome: Instituto do Studi del vincino oriente, 1967.

Carroll, Robert P. *When Prophecy Failed.* London: SCM Press, Ltd., 1979.

Childs, Brevard S. *The Book of Exodus, a Critical, Theological Commentary.* The Old Testament Library. Philadelphia: The Westminster Press, 1974.

Christensen, Duane L. *Transformations of the War Oracles in Old Testament Prophecy: Studies in the Oracles against the Nations.* Harvard Dissertations in Religion, 3. Missoula, Mont.: Scholars Press, 1975.

Clements, R. E. *Abraham and David: Genesis XV and Its Meaning in Israelite Tradition.* Studies in Biblical Theology, 2nd series, 5. London: SCM Press, Ltd., 1967.

Clements, R. E. *God and Temple.* Philadelphia: Fortress Press, 1965.

Clifford, Richard J. *The Cosmic Mountain in Canaan and the Old Testament.* Harvard Semitic Monographs, 4. Cambridge, Mass.: Harvard University Press, 1972.

Cogan, Morton. *Imperialism and Religion: Assyria, Judah, and Israel in the Eighth and Seventh Centuries B.C.E.* Missoula, Mont.: Scholars Press, 1974.

Coote, Robert B. *Amos among the Prophets: Composition and Theology.* Philadelphia: Fortress Press, 1981.

Craigie, Peter C. *The Problem of War in the Old Testament.* Grand Rapids, Mich.: William B. Eerdmans, 1978.

Cripps, Richard S. *A Critical & Exegetical Commentary on the Book of Amos.* London: SPCK, 1929.

Cross, Frank M., Jr. *Canaanite Myth and Hebrew Epic: Essays in the History of the Religion of Israel.* Cambridge, Mass.: Harvard University Press, 1973.

Davies, Eryl W. *Prophecy and Ethics: Isaiah and the Ethical Traditions of Israel.* Suppl. series 16. Sheffield, England: Journal for the Study of the Old Testament Press, 1981.

Day, John. *God's Conflict with the Dragon and the Sea: Echoes of a Canaanite Myth in the Old Testament.* New York: Cambridge University Press, 1985.

de Geus, C. H. J. *The Tribes of Israel: An Investigation into Some of the Presuppositions of Martin Noth's Amphictyony Hypothesis.* Amsterdam: Gorcum, Assen, 1976.

de Vaux, Roland. *Ancient Israel: Its Life and Institutions.* London: Darton, Longman & Todd, 1961.

Dothan, Trude. *The Philistines and Their Material Culture.* New Haven, Conn.: Yale University Press, 1982.

Eichrodt, Walter. *Theology of the Old Testament.* Vol. 1. Old Testament Library. Trans. John Baker. London: SCM Press, Ltd. 1961.

Engnell, Ivan. *Studies in Divine Kingship in the Ancient Near East.* New edition. Oxford: Basil Blackwell, 1967.

The Epic of Gilgamesh. Trans. N. K. Sandars. Baltimore: Penguin Books, 1960.

Frankfort, Henri. *Kingship and the Gods: A Study of Ancient Near Eastern Religion and the Integration of Society and Nature.* Chicago: University of Chicago Press, 1948.

Gadd, C. F. *Idea of Divine Rule in the Ancient Near East.* Lecture II. London: Oxford University Press, 1948.

Gevirtz, S. *Patterns in the Early Forms of Prophetic Speech.* Studies in Ancient Oriental Civilization, 32. Chicago: University of Chicago Press, 1963.

Gibson, J. C. L. *Canaanite Myths and Legends.* 2nd ed., rev. of 1st ed. by G. R. Driver. Edinburgh: T. & T. Clark, Ltd., 1977.

Glazer, Nathan. *American Judaism.* Chicago: University of Chicago Press, 1957.

Gottwald, Norman K. *All the Kingdoms of the Earth: Israelite Prophecy and International Relations in the Ancient Near East.* New York: Harper & Row, 1964.

Gottwald, Norman K. *The Hebrew Bible—A Socio-Literary Introduction.* Philadelphia: Fortress Press, 1985.

Gottwald, Norman K. *The Tribes of Yahweh: A Sociology of the Religion of Liberated Israel, 1250–1050 B.C.E.* Maryknoll, N.Y.: Orbis Books, 1979.

Gray, John. *I and II Kings, A Commentary.* Old Testament Library. 2nd ed. London: SCM Press, Ltd., 1970.

Gray, John. *The Legacy of Canaan: The Ras Shamra Texts and Their Relevance to the Old Testament.* 2nd rev. ed. VTSup 5. Leiden, The Netherlands: E. J. Brill, 1965.

Guillaume, A. *Prophecy and Divination among the Hebrews and Other Semites.* New York: Harper & Brothers, 1938.

Gunn, David, M. *The Story of King David: Genre and Interpretation.* Suppl. series 6. Sheffield, England: Journal for the Study of the Old Testament Press, 1978.

Gutiérrez, Gustavo. *A Theology of Liberation: History, Politics, and Salvation.* Trans. and eds. Sister Caridad Inda and John Eagleson. Maryknoll, N.Y.: Orbis Books, 1973.

Habel, Norman C. *Literary Criticism of the Old Testament.* Philadelphia: Fortress Press, 1971.

Haldar, A. *Associations of Cult Prophets among the Ancient Semites.* Uppsala: Almquist & Wiksells, 1945.

Halpern, Baruch. *The Constitution of the Monarchy in Israel.* Chico, Calif.: Scholars Press, 1981.

Hammershaimb, Erling. *The Book of Amos, a Commentary.* Trans. John Sturdy. Oxford: Basil Blackwell, 1970.

Harper, William Rainey. *A Critical and Exegetical Commentary on Amos and Hosea.* International Critical Commentary. New York: Charles Scribner's Sons, 1905.

Hayes, John H., ed. *Old Testament Form Criticism.* Trinity University Monograph Series in Religion, 2. San Antonio, Tex.: Trinity University Press, 1974.

Heaton, E. W. *The Old Testament Prophets.* Atlanta, Ga.: John Knox Press, 1977.

Heaton, E. W. *Solomon's New Men: The Emergence of Ancient Israel as a National State.* New York: Pica Press, 1974.

Henshaw, T. *The Latter Prophets.* London: George Allen & Unwin, Ltd., 1958.

Heschel, Abraham J. *The Prophets.* New York: The Jewish Publication Society of America, 1962.

Hillers, Delbert R. *Micah, a Commentary on the Book of the Prophet Micah.* Hermeneia Commentary. Eds. Paul D. Hanson and Loren Fisher. Philadelphia: Fortress Press, 1984.

Holladay, William L. *The Root Šûbh in the Old Testament.* Leiden, The Netherlands: E. J. Brill, 1958.

Hooke, S. H. *Babylonian and Assyrian Religion.* Norman: University of Oklahoma Press, 1963.

Hunter, A. Vanlier. *Seek the Lord! A Study of the Meaning and Function of the Exhortation in Amos, Hosea, Isaiah, Micah, and Zephaniah.* Baltimore: St. Mary's Seminary & University, 1982.

Hyatt, J. Philip. *Prophetic Religion.* New York: Abingdon Press, 1947.

Ishida, Tomoo. *The Royal Dynasties in Ancient Israel: A Study on the Formation and Development of Royal-Dynastic Ideology.* BZAW 142. New York: Walter de Gruyter, 1977.

Jacobi, W. *Die Ekstase der alttestamentlichen Propheten.* München: J. F. Bergmann, 1920.

Jacobsen, Thorkild. *The Treasures of Darkness: A History of Mesopotamian Religion.* New Haven, Conn.: Yale University Press, 1976.

Johnson, Aubrey R. *Sacral Kingship in Ancient Israel.* Cardiff: University of Wales, 1967.

Kaiser, Otto. *Isaiah 1–12, a Commentary.* The Old Testament Library. Philadelphia: The Westminster Press, 1972.

Kapelrud, Arvid S. *Central Ideas in Amos.* Oslo: W. Nygaard, 1956.

Klein, Ralph W. *Textual Criticism of the Old Testament: The Septuagint after Qumran.* Philadelphia: Fortress Press, 1974.

Klein, W. C. *The Psychological Pattern of Old Testament Prophecy.* Evanston, Ill.: Seabury-Western Theological Seminary, 1956.

Knight, H. *The Hebrew Prophetic Consciousness.* London: Lutterworth Press, 1947.

Koch, Klaus. *Amos: Untersucht mit den Methoden einer strukturalen Formgeschichte.* Alter Orient und Altes Testament, 30. Kevelaer, Federal Republic of Germany: Verlag Butzon; & Bercker/Neukirchen-Vluyn: Neukirchener Verlag, 1976. Part 1.

Koch, Klaus. *The Prophets: Volume One, the Assyrian Period.* Trans. Margaret Kohl. Philadelphia: Fortress Press, 1983.

Köhler, Ludwig. *Hebrew Man.* Trans. Peter R. Ackroyd. Nashville, Tenn.: Abingdon Press, 1956.

Köhler, Ludwig. *Old Testament Theology.* London: Lutterworth Press, 1957.

Kraeling, Emil G. *The Prophets.* New York: Rand McNally & Co., 1969.

Kraus, Hans-Joachim. *Worship in Israel: A Cultic History of the Old Testament.* Trans. G. Buswell. Richmond, Va.: John Knox Press, 1966.

Krentz, Edgar. *The Historical-Critical Method.* Philadelphia: Fortress Press, 1975.

Kuhl, Curt. *The Prophets of Israel.* Trans. Rudolf J. Ehrlich and J. P. Smith. London: Oliver & Boyd, 1960.

Lance, H. Darrell. *The Old Testament and the Archaeologist.* Philadelphia: Fortress Press, 1981.

Lang, Bernhard. *Monotheism and the Prophetic Minority.* The Social World of Biblical Antiquities. Sheffield, England: The Almond Press, 1983.

Lind, Millard C. *Yahweh Is a Warrior: The Theology of Warfare in Ancient Israel.* Scottdale, Pa./Kitchner, Ont.: Herald Press, 1980.

Lindblom, J. *Prophecy in Ancient Israel.* Philadelphia: Fortress Press, 1962.

Marsh, John. *Amos and Micah: Thus Saith the Lord.* Torch Bible Commentary. London: SCM Press, Ltd., 1959.

Mays, James L. *Amos, a Commentary.* Old Testament Library. Philadelphia: The Westminster Press, 1969.

Mays, James L. *Micah, a Commentary.* Old Testament Library. Philadelphia: The Westminster Press, 1976.

McKay, J. W. *Religion in Judah under the Assyrians, 732–609 B.C.* Studies in Biblical Theology, 2nd series, 26. London: SCM Press, Ltd., 1973.

Mendenhall, George. *The Tenth Generation: The Origin of the Biblical Tradition.* Baltimore: Johns Hopkins Press, 1973.

Mettinger, Tryggve N. D. *King and Messiah: The Civil and Sacral Legitimation of the Israelite Kings.* Lund, Sweden: C. W. K. Gleerup, 1976.

Mettinger, Tryggve N. D. *Solomonic State Officials: A Study of the Civil Government Officials of the Israelite Monarchy.* Lund, Sweden: C. W. K. Gleerup, 1971.

Miller, J. Maxwell. *The Old Testament and the Historian.* Philadelphia: Fortress Press, 1976.

Miller, J. Maxwell and John H. Hayes. *A History of Ancient Israel and Judah.* Philadelphia: The Westminster Press, 1986.

Miller, Patrick D., Jr. *The Divine Warrior in Early Israel.* Harvard Semitic Monographs, 5. Cambridge, Mass.: Harvard University Press, 1973.

Mowinckel, Sigmund. *He That Cometh.* Trans. G. W. Anderson. New York: Abingdon Press, 1954.

Mullen, E. Theodore, Jr. *The Divine Council in Canaanite and Early Hebrew Literature.* Harvard Semitic Monographs, 24. Ed. Frank M. Cross, Jr. Chico, Calif.: Scholars Press, 1980.

Myers, Jacob. *Hosea, Joel, Amos, Obadiah, Jonah.* Layman's Bible Commentary, 14. Richmond, Va.: John Knox Press, 1959.

Myers, Jacob M. *II Chronicles.* The Anchor Bible. New York: Doubleday & Company, Inc., 1965.

Nicholson, Ernest W. *The Book of the Prophet Jeremiah, Chapters 1–25.* Cambridge Commentary. Cambridge: Cambridge University Press, 1973.

Noth, Martin. *Numbers, a Commentary.* The Old Testament Library. Trans. James D. Martin. Philadelphia: The Westminster Press, 1968.

Östborn, Gunnar. *Yahweh and Baal: Studies in the Book of Hosea and Related Documents.* Lund, Sweden: C. W. K. Gleerup, 1956.

Phillips, Anthony. *Ancient Israel's Criminal Law: A New Approach to the Decalogue.* Oxford: Basil Blackwell, 1970.

Pope, Marvin H. *El in the Ugaritic Texts.* VTSup 2. Leiden, The Netherlands: E. J. Brill, 1955.

Pope, Marvin H. *Song of Songs.* The Anchor Bible. New York: Doubleday & Company, Inc., 1977.

Pritchard, James B., ed. *Ancient Near Eastern Texts Relating to the Old Testament.* 2nd ed. Princeton: Princeton University Press, 1955.

Pritchard, James B., ed. *Ancient Near Eastern Texts, Supplementary Texts and Pictures Relating to the Old Testament.* Princeton: Princeton University Press, 1969.

Pritchard, James B., ed. *The Ancient Near East in Pictures Relating to the Old Testament.* Princeton: Princeton University Press, 1954.

von Rad, Gerhard. *Deuteronomy, a Commentary.* The Old Testament Library. Trans. Dorothea Barton. Philadelphia: The Westminster Press, 1966.

von Rad, Gerhard. *Old Testament Theology, Vol. II, the Theology of Israel's Prophetic Traditions.* Trans. D. M. G. Stalker. New York: Harper & Row, 1965.

von Rad, Gerhard. *The Problem of the Hexateuch and Other Essays.* Trans. E. W. Trueman Dicken. London: Oliver & Boyd, 1966.

Rast, Walter E. *Tradition History and the Old Testament.* Philadelphia: Fortress Press, 1972.

Robertson, David. *The Old Testament and the Literary Critic.* Philadelphia: Fortress Press, 1971.

Robinson, H. Wheeler. *Inspiration and Revelation in the Old Testament.* Oxford: Clarendon Press, 1946.

Rudolph, Wilhelm. *Joel–Amos–Obadiah–Jona.* Kommentar zum Alten Testament. Gütersloh, Federal Republic of Germany: Gerd Mohn, 1971.

Sanders, James A. *Canon and Community: A Guide to Canonical Criticism.* Philadelphia: Fortress Press, 1984.

Van Seters, John. *In Search of History: Historiography in the Ancient World and the Origins of Biblical History.* New Haven, Conn.: Yale University Press, 1983.

Skinner, John. *Prophecy and Religion.* Cambridge: Cambridge University Press, 1922.

Smend, Rudolf. *Yahweh War and Tribal Confederation: Reflections upon Israel's Earliest History.* Trans. Max Gray Rogers. New York: Abingdon Press, 1970.

Soggins, J. Alberto. *A History of Ancient Israel*. Philadelphia: The West-minster Press, 1984.

Steindorff, George and Keith C. Seele. *When Egypt Ruled the East*. Chicago: University of Chicago Press, 1942.

Tamez, Elsa. *Bible of the Oppressed*. Trans. Matthew J. O'Connell. Mary-knoll, N.Y.: Orbis Books, 1982.

Thomas, D. Winton, ed. *Documents from Old Testament Times*. New York: Thomas Nelson & Sons, Ltd., 1958.

Tucker, Gene M. *Form Criticism of the Old Testament*. Philadelphia: Fortress Press, 1971.

Ward, James M. *Amos & Isaiah: Prophet of the Word of God*. New York: Abingdon Press, 1969.

Watts, John D. W. *Vision and Prophecy in Amos*. Leiden, The Netherlands: E. J. Brill, 1958.

Weinfeld, Moshe. *Deuteronomy and the Deuteronomic School*. Oxford: Clar-endon Press, 1972.

Weiser, Artur. *Die Profetie des Amos*. BZAW 53. Giessen, Germany: Alfred Topelmann, 1929.

Weiser, Artur. *The Psalms: A Commentary*. Old Testament Library. Trans. Herbert Hartwell. Philadelphia: The Westminster Press, 1962.

Welch, A. C. *Prophet and Priest in Old Israel*. Oxford: Basil Blackwell, 1936.

Westermann, Claus. *Basic Forms of Prophetic Speech*. Trans. Hugh Clayton White. Philadelphia: The Westminster Press, 1967.

Whitelam, Keith W. *The Just King: Monarchical Judicial Authority in Ancient Israel*. Suppl. series 12. Sheffield, England: Journal for the Study of the Old Testament Press, 1979.

Whitley, C. F. *The Prophetic Achievement*. London: A. R. Mowbray & Co., 1963.

Whybray, R. N. *The Heavenly Counsellor in Isaiah x1 13–14: A Study of the Sources of the Theology of Deutero–Isaiah*. Cambridge: Cambridge University Press, 1971.

Whybray, R. N. *The Succession Narrative: A Study of II Sam. 9–20; I Kings 1 and 2*. Studies in Biblical Theology, 2nd series, 9. London: SCM Press, Ltd., 1968.

Widengren, G. *Sacrales Königtum im Alten Testament und im Judentum*. Stuttgart: Verlag W. Kohlhammer GmbH, 1955.

Wilson, Robert R. *Sociological Approaches to the Old Testament*. Philadel-phia: Fortress Press, 1984.

Wolff, Hans Walter. *Hosea, A Commentary on the Book of the Prophet Hosea*. Hermeneia Commentary. Trans. Gary Stansell. Ed. Paul D. Hanson. Philadelphia: Fortress Press, 1974.

Wolff, Hans Walter. *Micah the Prophet*. Trans. Ralph D. Gehrke. Phila-delphia: Fortress Press, 1981.

Wolff, Hans Walter. *Joel and Amos*. Hermeneia Commentary. Trans. Waldemar Janzen, S. Dean McBride, Jr., and Charles A. Muenchow. Philadelphia: Fortress Press, 1977.

Yadin, Yigael. *The Art of Warfare in Biblical Lands*. Vol. II. New York: McGraw-Hill Book Company, Inc., 1963.

Yadin, Yigael et al. *Hazor II: An Account of the Second Season of Excavation*. Jerusalem: Hebrew University (Magnes Press), 1960

Index of Biblical Citations

229

Author and Subject Index